Springer Biographies

The books published in the Springer Biographies tell of the life and work of scholars, innovators, and pioneers in all fields of learning and throughout the ages. Prominent scientists and philosophers will feature, but so too will lesser known personalities whose significant contributions deserve greater recognition and whose remarkable life stories will stir and motivate readers. Authored by historians and other academic writers, the volumes describe and analyse the main achievements of their subjects in manner accessible to nonspecialists, interweaving these with salient aspects of the protagonists' personal lives. Autobiographies and memoirs also fall into the scope of the series.

Tianxin Cai
Mathematical Legends

From Thales to Erdős

 Springer

Tianxin Cai
School of Mathematics
Zhejiang University
Hangzhou, Zhejiang, China

Translated by
Tyler Ross
New York, USA

ISSN 2365-0613   ISSN 2365-0621 (electronic)
Springer Biographies
ISBN 978-981-96-2402-7   ISBN 978-981-96-2403-4 (eBook)
https://doi.org/10.1007/978-981-96-2403-4

The Work already has been published in 2022 in Chinese language by The Commercial Press with the following title: Shu Xue Chuan Qi. Therefore, notwithstanding the above, the Author grants these rights to the Publisher only for editions in E

© The Editor(s) (if applicable) and The Author(s), under exclusive license to Springer Nature Singapore Pte Ltd. 2025

This work is subject to copyright. All rights are solely and exclusively licensed by the Publisher, whether the whole or part of the material is concerned, specifically the rights of reprinting, reuse of illustrations, recitation, broadcasting, reproduction on microfilms or in any other physical way, and transmission or information storage and retrieval, electronic adaptation, computer software, or by similar or dissimilar methodology now known or hereafter developed.
The use of general descriptive names, registered names, trademarks, service marks, etc. in this publication does not imply, even in the absence of a specific statement, that such names are exempt from the relevant protective laws and regulations and therefore free for general use.
The publisher, the authors and the editors are safe to assume that the advice and information in this book are believed to be true and accurate at the date of publication. Neither the publisher nor the authors or the editors give a warranty, expressed or implied, with respect to the material contained herein or for any errors or omissions that may have been made. The publisher remains neutral with regard to jurisdictional claims in published maps and institutional affiliations.

This Springer imprint is published by the registered company Springer Nature Singapore Pte Ltd.
The registered company address is: 152 Beach Road, #21-01/04 Gateway East, Singapore 189721, Singapore

If disposing of this product, please recycle the paper.

Preface

The great German poet Goethe, who sustained a lifelong interest in the study of natural phenomena, once wrote: "The history of science is science itself." This book is not about the history of mathematics, but it discusses distinctive figures in the history of mathematics, revealing all manners of strange treasures, bright flowers, and secret passions within the mathematical kingdom.

Some of these great mathematicians also made outstanding contributions to the humanities, such as Thales, Khayyam, Descartes, Pascal, Leibniz, and von Neumann; others lived through legendary personal experiences: Archimedes, Qin Jiushao, Erdős, Shiing-Shen Chern, and Hua Luogeng. Some of them were thinkers, writers, poets, and economists, and some were physicists, astronomers, government officials, soldiers, and monks.

During more than 40 years of mathematical practice and theoretical training, the author has become more and more aware of how insignificant is any person in the face of the vast ocean of mathematics, even more so in the abstract twenty-first century. Fortunately, I have taken advantage of various opportunities to visit the home countries of every character appearing in this book, giving me a clearer grasp on the trajectories of their lives.

Many people have heard of the sentence inscribed upon the entrance to Plato's Academy: "Let no one ignorant of geometry enter here." At that time, geometry was more or less synonymous with mathematics as a whole. Fewer people realize that there was another inscription upon its exit: "Only those who understand philosophy are fit to govern a country." Several of the mathematicians mentioned in this book were also accomplished philosophers.

Moreover, there is a natural connection between mathematics and the humanistic spirit. The achievements and teachings of our mathematical ancestors are a great source of warmth, as the fifth-century Byzantine philosopher Proclus wrote:

> This, therefore, is mathematics: she reminds you of the invisible form of the soul; she gives light to her own discoveries; she awakens the mind and purifies the intellect; she brings light to our intrinsic ideas; she abolishes oblivion and ignorance which are ours by birth.

This book also contains two special articles, exploring respectively the relationship between mathematicians and politicians, and mathematicians and poets, each touching upon quite a few characters. Another is an interview with the physicist Chen-Ning Yang, who is now over 100 years old. His father was a number theorist, and he himself loved mathematics and had a special feeling for it.

"Genius is diligence." This sentence is more concise and powerful than the famous quotation from Thomas Edison, "Genius is one percent inspiration, and 99% perspiration." It also summarizes successfully the experiences of the protagonists of this book, although they are fewer than half of the characters in the Chinese book of the same name. This saying comes to us from the eighteenth-century German physicist and satirist Georg Christoph Lichtenberg, who invented xerography technology, the earliest of the famous professors at the University of Göttingen.

Carl Friedrich Gauss, the "prince of mathematics," was another alumnus of the University of Göttingen, as was his lifelong friend, the Hungarian mathematician Farkas Bolyai, who once wrote that many things seem to have a time when they are discovered in many places all at the same time, just as violets are seen blooming everywhere in spring. We look forward to a bright future for mathematics, with flowers blooming more beautifully in countries all over the world.

Contents

1 Thales of Miletus, First of The Seven Sages — 1

2 Archimedes: The God of Mathematics — 15

3 The World of Omar Khayyam — 35

4 Qin Jiushao, Daogu Bridge, and the *Mathematical Treatise in Nine Sections* — 55

5 The Reclusive Frenchmen: Descartes and Pascal — 73

6 Leibniz: Unattainable Heights — 89

7 John von Neumann, Who Made the World a Better Place — 107

8 Paul Erdős: A Narrowly Missed Opportunity — 139

9 Mathematicians and Poets — 153

10 Mathematicians and Political Leaders — 163

11 Hua Luogeng and Shiing-Shen Chern: Two Contemporary Chinese Masters — 175

12 "My Life Can Be Said to Form a Circle"—*An Interview with Nobel Laureate Professor Chen-Ning Yang* — 205

1

Thales of Miletus, First of The Seven Sages

Let no one ignorant of geometry enter here.
—*Plato*

Thales of Miletus

The history of human civilization is rife with coincidences one after another. For one example, on April 23, 1616, William Shakespeare, the greatest writer of the English-speaking world, and Miguel de Cervantes, the greatest writer of the Spanish-speaking world, both died on the same day (this day later became World Book Day). Moreover, the Italian painter Leonardo da Vinci, regarded as the most perfect representative of the Renaissance, was born on April 15, 1452, according to the Julian calendar, which is also April 23 according to the Gregorian calendar. Another concurrence: the greatest Italian scientist Galileo Galilei died on January 8, 1642, and within a year the greatest British scientist Isaac Newton was born. In earlier times, a string of mathematicians and philosophers emerged in ancient Greece in great numbers, just like the writers and artists in Italy during the Renaissance.

In the year 1266, the second year after the great poet Dante Alighieri was born in Florence, this city gave birth to Giotto di Bondone, the most outstanding artist of that century. According to the British art historian Sir Ernst Gombrich, artists were regarded prior to Giotto in the same way as excellent

Likeness of Thales of Miletus

carpenters or tailors, and they often did not even sign their works. But after Giotto, the history of art has ever since been inseparable from the history of artists.

By way of contrast, mathematicians have been much more smiled upon by history. The first mathematician whose name is known is Thales of Miletus (ca. 625–527 BCE), of ancient Greece, living nearly 19 centuries earlier than Giotto. Thales was born in Miletus in Asia Minor on the eastern shores of the Aegean Sea (today the western coast of the Asian portion of Türkiye), 1 of the 12 city states of Ionia. The Ionians comprised a tribe that had originally been scattered throughout Greece before migrating to Asia Minor and forming a community, so that that area too came to be called Ionia. They became rich and powerful as traders and subsequently formed an alliance among themselves.

As a philosopher of the pre-Socratic era, Thales is known as the first of The Seven Sages of Greece, the others being Solon of Athens, Chilon of Sparta,

Cleobulus of Rhodes, Periander of Corinth, Pittacus of Lesbos, and Bias of Priene, also in Asia Minor. Due to the long march of history and the fact that at that time people could still only transmit their ideas as oral tradition, the lives and deeds of the five sages other than Thales and Solon cannot be verified. We know only that they were variously statesmen and rulers, from each of whom only a saying or two has been passed down.

For example, Bias is associated with the phrase, "the majority of people are wicked," Periander with "think before acting," Pittacus with "know thine opportunity," and Cleobulus with "moderation is the best thing," which is similar to the famous dictum of Solon to avoid extremes and also to Chinese Confucianism. So too, the wisdom of Periander recalls the Chinese idiom, "think twice before acting [三思而后行]," first spoken by Ji Wenzi, a senior official of the state of Lu, although Confucius, also from the state of Lu, disapproved of his vacillating over gains and losses. Thales and Solon are the source of many maxims that have made it down to subsequent generations, among which I particularly admire "know thyself," attributed to Thales, and "speech is the mirror of action," attributed to Solon.

Miletus was the largest among the eastern cities of Greece at that time and took its name from a region in Crete. Indeed, most of the residents of Miletus were immigrants from Crete. In this city, merchant rule had replaced clan aristocracy, resulting in a culture of freer and more open minds that led to the production of many famous figures in literature, art, science, and philosophy. The blind poet Homer, and later the historian Herodotus, both hailed from Ionia. Thales is said to have engaged in business in his early years, to have traveled to Egypt and Babylonia, and to have learned and mastered mathematics and astronomy. In addition to these two fields, Thales later carried out research into physics, engineering, and philosophy.

Only Thales among the Seven Sages was a knowledgeable scholar. He founded the Ionian School of philosophers and sought to eradicate religion in favor of the pursuit of truth in natural phenomena. Thales believed that life and movement permeate everything and viewed water as the origin of all things. In his youth, Thales took advantage of his business career to sustain extensive contact with society. Here we relate an anecdote relating Thales and water: once, when he was using mules to transport salt, one of them slipped into a stream, causing some of the salt to dissolve and thereby noticeably lightening its load. As a result, the mule deliberately rolled into the water at every stream it came to. In order to rid the beast of this bad habit, Thales loaded a sponge onto its back so that instead the weight it carried would be doubled upon contact with water. The mule never dared again to get up to its old tricks from that point on.

Thales is said to have gained proficiency in mathematical techniques during his time in Egypt. He once calculated the height of a pyramid via the ratio of its shadow under the sun with that cast by a pole. According to this widely spread story, Thales took the opportunity of a sunny day to place a pole vertically in the ground. He waited until the length of its shadow was equal to its height (according to some versions he used instead his own height and silhouette) to measure the length of the shadow of the pyramid, which gave its height. However, since the base of a pyramid is large, and not a point, such a measurement would only be possible at special angles of sunlight. A version of the story that accounts for this has it that Thales placed the pole at the end of the shadow of the pyramid, causing two similar triangles to be formed by the projection of the shadows. Then the ratio of the height of the pyramid to the height of the pole is identical to the ratio of the lengths of their shadows.

Thales Through the Eyes of the Philosophers

Although the name of Thales has made its mark on history, we nevertheless know very little about his life. Fortunately, a few anecdotes concerning him are mentioned in the works of later philosophers and writers, from which we can piece together an understanding of his temperament and personality. This, perhaps, is the earliest story in mathematics. It is regrettable that although some famous mathematicians also appeared in ancient China, no such cultural atmosphere embraced them, and scholars of the humanities rarely paid attention to the work of scientists. There are a few exceptions: Zhuangzi made note in the *Tianxia* (*Under Heaven*) of the concept of infinity as expounded by the famous scholar Hui Shi, and in the mathematical treatise *Zhoubi Suanjing*, it is mentioned that the Duke of Zhou discussed the Pythagorean theorem with the astrologer Shang Gao.

We turn first to Plato, who was both a philosopher and a mathematician. Legend has it that above the entrance of the academy he founded was inscribed the dictum, "Let no one ignorant of geometry enter here," and at its back door, "Only those who understand philosophy are fit to govern a country." Among his important works is the dialogue *Theaetetus*, named after another student (ca. 417–369 BCE) of Socrates, somewhat older than Plato, who was present with Plato at the scene of the death of their teacher. Theaetetus too was both a mathematician and a philosopher, the founder of solid geometry, and the main interlocutor of both this dialogue and another, *The Sophist*.

In the *Theaetetus*, which serves as a kind of memorial to his teacher and senior, Plato discusses the nature of knowledge. Socrates proposes three answers to the question "What is knowledge?" to young Theaetetus: perception, true judgment, and true judgment justified by reason, each of which he ultimately rejects, since for the seeker of knowledge the most important thing is the process. This dialogue also contains an anecdote concerning Thales, in which the astronomer became so distracted in looking up at the stars that he tripped and fell into a well, provoking the laughter of a beautiful Thracian girl who asked him how he could hope to know the things in heaven when he cannot even see the ground in front of his eyes. Thales made no response to this, but he was however strung by a different question posed to him by Solon, archon of Athens.

This question related to the fact that Thales was probably the first of many sages and scholars to live out his entire life in solitude. According to Plutarch, a Roman biographer of the first century (more than six centuries after the death of Thales), Solon, who was 14 years older than Thales, once paid a visit to the latter in Miletus. Solon was a politician, reformer, and legislator who served as archon or chief magistrate in Athens in 594 BCE. He was also a successful businessman who liked to travel to famous mountains and rivers and to survey social customs. He was even accomplished as a poet, known as the first poet of Athens. Although his poetry consisted mostly of praise for the city-state of Athens and its laws, he also condemned the greed, tyranny, and cruelty of the nobles and recorded his firm belief that morality is more valuable than wealth, once writing in a poem:

> Some wicked men are rich, some good are poor;
>   We will not change our virtue for their store:
> Virtue's a thing that none can take away,
> But money changes owners all the day.[1]

---

[1] Tr. John Dryden.

Relief portrait of Solon in the United States House of Representatives

From this it is clear that Solon was a highly individual figure. Thales was known for the maxim that excessive stability brings only disaster, quite at odds with the injunction of Solon: "avoid extremes," and sure enough, in the course of the conversation, there emerged another point of difference between them—Solon ventured to ask Thales why he had never married, which led Thales to become visibly unhappy and make no answer.

A few days later, Solon, who was a man of rich emotions, a poet and a lover of travel, received news that a young man who had recently and unfortunately died in Athens may have been his son, which left him devastated. At this point Thales appeared with a smile and told him that the story had been entirely fictitious but illustrated the reason he had never wanted to marry and have children: he was too afraid to face the pain of losing a loved one. Allegedly his mother too had urged him to marry when he was middle aged, to which Thales replied that the time had not come; later, when Thales was already an old man, his elderly mother pressed him again to get married at last, to which at this time he replied that the time had already past.

Plutarch, who was mentioned above, was an author of biographies whose works were very popular during the Renaissance. The French writer Michel de Montaigne praised him highly, and Shakespeare based many of his plays on the works of Plutarch. Each of his biographies was followed by a commentary. About Thales, Plutarch wrote that "… it is irrational and ignoble to renounce the acquisition of what we want for fear of losing it; … we must be fortified not by poverty against deprivation of worldly goods, nor by friendliness against loss of friends, nor by childlessness against loss of children, but by reason against all adversities."[2]

The philosopher Aristotle, a disciple of Plato active nearly three centuries after Thales, related another story concerning him in his *Metaphysics*. According to this story, Thales discerned one year on the basis of his knowledge of agriculture and meteorology that the following year would produce a bumper crop of olives and raised funds to rent in advance all the olive presses in the area at low prices. As he expected, the subsequent demand for presses outpaced the supply, and he was able to rent them out again at a high price and amass in this way a sizable fortune. But he did not do this for the sake of riches, but only so as to be able to reply to the question frequently posed to him: "if you really are so smart, why haven't you gotten rich?" Indeed, he continued to insist that knowledge is better than wealth.

Thales the Versatile

The first named historian of mathematics, Eudemus of Rhodes, who lived around the second half of the fourth century BCE, was a disciple of Aristotle who wrote works on the history of arithmetic, geometry, and astronomy; unfortunately, all of them have been lost. He also contributed to the editing of the complete works of his mentor, Aristotle. Fortunately, we do have a *Summary of Eudemus*, based mainly on his *History of Geometry*, written by Proclus (410–485 CE) when he was preparing a commentary on *The Elements* by Euclid. Eudemus wrote in this book that it was Thales who introduced the study of geometry from Egypt to Greece and that Thales himself moreover had discovered many propositions and directed his students toward the study of fundamental principles from which other propositions could be derived.

Proclus the Successor, last director of the Platonic Academy, famously said that wherever there is number, there is beauty. It is precisely due to the records of Proclus that we know that Thales proved five theorems in plane geometry, including the theorem known today as Thales's theorem, all of which appear in middle school mathematics textbooks to this day. Thales proved the following: the diameter of a circle divides it into two equal parts; the two angles at the base of an isosceles triangle are equal; the opposite angles at point of intersection of two straight lines are equal; if two triangles have two angles and one side that are equal, then the two triangles are congruent.

---

[2] Tr. Bernadette Perrin.

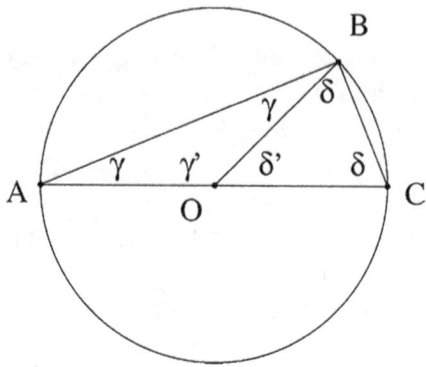

Proof of Thales's Theorem

Of course, the most valuable mathematical contribution due to Thales is the theorem that now bears his name, which states that the angle in the semicircle in any circle is a right angle. This theorem, which appears in Euclid's *Elements* as Proposition 31 of Book III, is the first theorem in the history of mathematics named after a mathematician. The angle in the semicircle here refers to the angle formed by taking the vertex anywhere along the circumference of the circle and joining it by straight lines to the two endpoints of any diameter of the circle. The proof uses one of the equivalent formulations of the parallel postulate, specifically that the angles in a triangle sum to two right angles. We present the details as follows:

> Let AC be the diameter bisecting the circle and B the vertex of the angle in the semicircle. We wish to show that the angle ABC is a right angle. Join the center O of the circle to point B. From the fact that the two angles at the base of an isosceles triangle are equal, we see that the angle OAB is equal to the angle ABO and similarly that the angle OCB is equal to the angle CBO. It follows that the angle ABC is given by the sum of the angles OAB and OCB, and since the sum of the angles in the triangle ABC must equal two right angles, we find that the angle ABC is equal to a right angle.

Thales also achieved extraordinary things beyond the scope of mathematics. He famously asserted that water is the greatest thing, and the first principle of everything, following upon the observation that water evaporates in sunlight, its mist rises to form clouds, and the clouds turn into rain. Intriguingly, Thales included metals in his category of water, probably because they can be melted. He furthermore believed that the earth has the form of a disk floating atop water and that earthquakes were vibrations caused by the drifting currents. Although this view was later disproved, it shows that Thales dared to contemplate the true face of nature and establish his own ideological system, for which reason he is recognized as the originator of Greek philosophy.

In fact, Thales was the first to raise the philosophical question of the origin of all things and attempt to answer it. He pioneered the spirit of rationalism, the tradition of materialism, and the principle of universality. With respect to his theology, Thales was a polytheist, who believed the world to be filled with various gods, but he nevertheless sought reasons and explanations for things in nature itself, rather than in the elusive character of gods with human forms. This is the fundamental feature defining Thales as a thinker. In his later years, Thales recruited students, establishing the Ionian School of philosophy.

In physics, Thales is also credited with the discovery that friction produces static electricity in amber. He studied astronomy extensively, estimated the sizes of the sun and the moon, identified the Ursa Minor constellation pointing out its utility for navigation, and established for the first time the length of a year as 365 days. A story concerning his astronomical skills comes to us from Herodotus, who died 2 years before the birth of Plato and is regarded as the father of history on the basis of his work, the *Histories*. This is the first prose work to be handed down intact in the West, so that Herodotus can also be viewed as the founding figure of Western literature and an outstanding example of humanism.

Herodotus records that Thales accurately predicted a solar eclipse in the year 585 BCE, which brought a halt to a war. The conflict was between the Lydians, led by Alyattes, on the one side, and the Medes, led by Cyaxares, on the other. They had been fighting for five consecutive years without any definitive victory or defeat on either side, but only the loss of lives and corpses everywhere. Thales anticipated the coming solar eclipse with the threat that it would be a warning from the gods of their opposition to war. As a result, when the eclipse took place as predicted in the middle of a fierce battle, and day turned into night, the soldiers became very frightened; they remembered his warning, stopped fighting, and made peace. This was the earliest clearly recorded solar eclipse in the West, the date being May 28th.

As for the method that Thales used to predict it, later scholars believe that he may have known of the saros cycle[3] discovered by the ancient Babylonians, although some modern scholars believe that Thales did not in fact possess sufficient knowledge to accurately predict the time and location of a solar eclipse.

---

[3] The saros is an astronomical term referring to the periodicity of solar and lunar eclipses; its length is approximately 6585.32 solar days, equivalently 18 years and either 10.3 or 11.3 days (depending on whether 5 or 4 leap years appear in the cycle). During each saros cycle, there are approximately 43 solar eclipses and 28 lunar eclipses.

Eudemus thought that Thales already knew that the four seasons as separated by the vernal equinox, summer solstice, autumnal equinox, and winter solstice were of unequal length. The interaction between Thales and Solon may be the earliest friendship between a mathematician and a politician and a mathematician and a poet in history.

The Students and Legacy of Thales

More important than the specific content of Thales's theorem is the fact that Thales introduced the idea of proofs for propositions. Thales pioneered this concept, which marks the evolution in human understanding of objective things from the perceptual to the rational, an extraordinary leap in the history of mathematics that would be carried much further forward centuries later by Euclid with the composition of his *Elements*. Thales furthermore relied in his proofs upon some axioms or basic propositions the proof of which was subsequently confirmed. Although we do not have any primary sources exhibiting these achievements, the records of them above have been passed down to the present. All of this earns Thales the title of the first mathematician and the originator of demonstrative geometry.

Thales maintained a humorous and philosophical outlook. His prescription for living an upright life was to "avoid doing what you would blame others for doing," identical to the dictum of Confucius: "do not do to others what you do not wish for yourself" in *The Analects: Yan Yuan*. When someone asked him what is the strangest thing he had ever seen, Thales replied "A long-lived tyrant." When he was asked what he wished to gain from making a discovery, he who had never received any reward said "do not tell others that it is your discovery but that it is mine; this is the highest reward."

The theories and ideas of Thales had a huge influence. He led the way toward rational exploration of the world and played the role of the first scientist and mathematician worthy of these names. Following his lead, the Greeks cast of the shackles of superstition and embarked upon the conscious exploration of the mysteries of numbers, shapes, and even the universe, uncovering the original secrets of nature. After centuries of hard work, mathematics underwent a transition from its concrete, experimental stage to its abstract, theoretical stage, gradually taking the form of an independent deductive doctrine that ultimately facilitated the prosperity of Greek science, art, and philosophy and went on to influence Europe and the entire world.

Among the students of Thales, the most accomplished were Anaximander and Anaximenes. Anaximander (ca. 610–545 BCE) was born in Miletus and served as a leader of a Milesian colony near the Black Sea. He believed that the world is not made of water but rather some special fundamental form that is not readily known but which underlies the four elements: earth, air, water, and fire. This substance splits via its motion the opposites of coldness and heat, dryness and wetness, thereby creating all things. The world both comes from it and returns to it. Anaximander was the first of the philosophers to write down his ideas in prose, in contrast to the verse of Homer and Hesiod.

Anaximander employed a method of reductio ad absurdum to conclude that humans evolved from fish of the sea and more generally that higher animals evolved from lower ones. He proposed the important cosmological view that the earth is a freely floating cylinder at the center of the universe, with the sun, moon, and stars arranged in a ring around it. This view of the universe survived for more than 2000 years until the emergence of the Copernican theory. Anaximander was said to have been good at acting, with a flair for dramatic costumes and speech, and he led an envoy to Sparta, where he demonstrated two of his great inventions: the sundial and the first map of the world. Unfortunately, his book *On Nature* has been lost.

Anaximenes (ca. 586–526 BCE) held a different view. He believed that the world is composed of air, the condensation and dispersion of which produce the various forms of matter; like his two predecessors in the Milesian school, his was a monistic philosophy. Where the Egyptians and Babylonians had called upon the gods to explain the formation and nature of the world, the Milesian philosophers gave naturalistic interpretations. Anaximenes made the following assumption: gas is an omnipotent substance, capable of spontaneously entering into our souls and enabling the control of our bodies.

It is said that Anaximenes had thousands of disciples. According to one legend, he once asked the students in one of his lectures to abandon their notetaking and listen carefully, promising to disseminate his remarks later; instead he gave them a blank sheet and asked that they reproduce what they had heard, but only Pythagoras, who was passing by and stopped in to listen, remembered everything. Of course, it is difficult to say whether such a story is true or false, but it is rich in philosophy and invokes the art of teaching and administrating, namely, to make the students or employees responsible for themselves in learning and mastering practical skills.

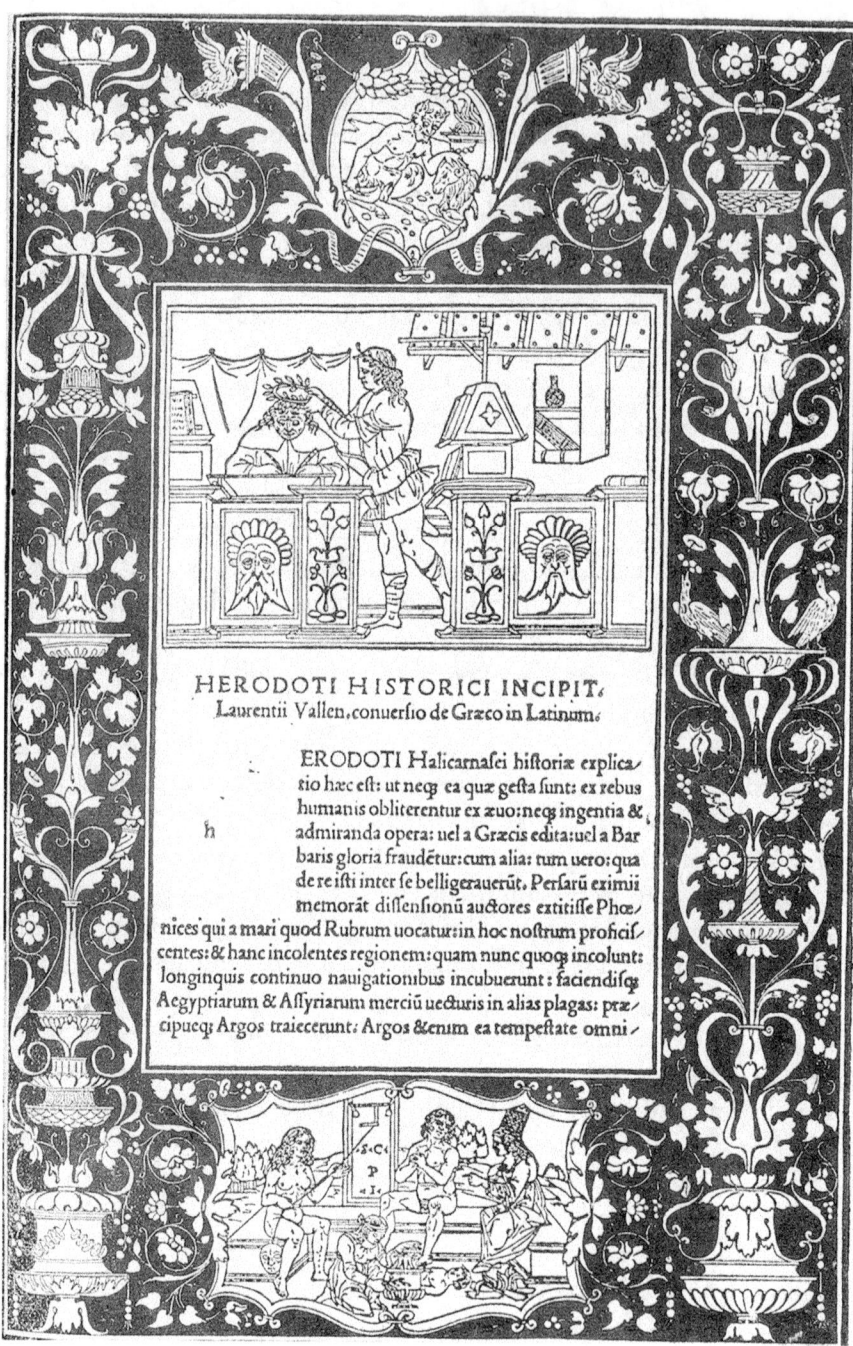

Latin edition of the *Histories* of Herodotus (1494, Venice)

Anaximander once described infinite suffering as the raw material of nature, which Anaximenes supported with an example of wool being compressed to make rugs.

The historian, writer, and traveler Hecataeus of Miletus (ca. 550–476 BCE) was also included among the disciples of Thales by some scholars. Hecataeus not only wrote the earliest travel notes in a beautiful and concise style, having traveled extensively through the Persian empire, but was also a pioneer in geography and ethnography. He famously observed that "Egypt is the gift of the Nile," a name that survived among later generations. But considering the years of his birth and the year in which Thales died, it is impossible that Hecataeus was a student of the latter. In any case, he was a Milesian during the period of Persian rule and the first known historian of ancient Greece. Herodotus, who was born shortly before the death of Hecataeus, adopted and modified his maxim, saying: "Geometry is the gift of the Nile."

West Brook, Hangzhou, 2022

2

Archimedes: The God of Mathematics

God is ever doing geometry.
—Plato

Ancient Syracuse

Archimedes was born in 287 BC in Syracuse, a port city in the southeastern corner of Sicily, the largest island in the Mediterranean Sea. The year of his birth is based on the year of his death and the presumed length of his life, which was given by the poet and scholar John Tzetzes of Constantinople (now Istanbul), an interesting figure in his own right, who was regarded as the model of a perfect pedant. Tzetzes was the son of an Iberian (Georgian) mother and served in his youth as secretary to a provincial governor before earning his living in later life as a teacher and writer. In his most influential work, usually referred to as the *Chiliades* (*Thousands*) but originally entitled *Book of Histories*, consisting of more than 12,000 lines of verse and quoting more than 400 authors, Tzetzes writes that "Archimedes the Wise was a Syracusan, a famous builder of machines, who spent his life studying geometry, and lived to the age of seventy-five years."

There had earlier been a biography of Archimedes written by his friend Heracleides (not to be confused with the better known sixth-century BCE philosopher Heraclitus, nor with the fourth-century BCE astronomer Heraclides, a student of Plato who later served as an administrator at his Academy, first suggested the rotation of Earth about its axis and believed that Mercury and Venus revolve about the sun). This biography is mentioned

Cathedral in Sicily, with geometry patterns along the outer wall (photograph by the author)

more than once by the sixth-century mathematical commentator Eutocius, but unfortunately it was later lost. The picture we have of the life of Archimedes, like that of Thales of Miletus, comes to us scattered across various ancient texts.

The ancient Greeks regarded themselves as divided into four tribes: the Achaeans (identified with the Mycenaens by some historians), the Ionians, the Dorians, and the Aeolians. The inhabitants of Syracuse were of Dorian descent, whereas Catania, slightly to the north, was inhabited by Ionians; among them was the philosopher Thales, who was credited with the founding of the Ionian school of philosophy. The Aeolians lived across the water in the southernmost tip of the Apennine peninsula. Slightly further to the north in Tarentum, home to the Pythagorean Academy, lived the Achaeans, who spoke a different dialect. The Dorians are first recorded in Homer's epic poem *The Odyssey*, according to which they lived on the island of Crete. Tracing backward their roots, the Dorians most likely came from the northern Balkans and later migrated to the Peloponnese, Rhodes, Crete, and the eastern part of Sicily. The Dorians in Sicily had immigrated for the most part from Corinth, at the border between the Peloponnese peninsula with mainland Greece.

The Syracusans had established an empire about a century before the birth of Archimedes, expanding their territory northward into southern Italy and waging three wars against Carthage, the capital city in what is now Tunisia in North Africa established by the Phoenicians, who came from the eastern

2 Archimedes: The God of Mathematics

The island of Ortygia in Syracuse, where Archimedes fought against the Romans

shores of the Mediterranean. But this Syracusan Empire suddenly collapsed 2 years before Archimedes was born. Indeed, by this time the heyday of ancient Greece had passed already, and the economic and cultural center had shifted to Alexandria, a Mediterranean port city in northern Egypt; at the same time, the nascent Roman empire in the Apennines was expanding in power. So Archimedes grew up in an era of transition between old and new, and his home city of Syracuse became the site of struggles between many forces.

Archimedes came from an aristocratic background: his father Phidias was an astronomer, with the same name, although not related to the earlier great sculptor, painter, and architect involved in the construction of the Parthenon on the Acropolis, from which some commentators have inferred that his grandfather was an artist or at least a lover of the arts. What is clear in any case is that Archimedes inherited from his father a love for thought and study. At around the age of 10, his father sent him to study in Alexandria, at that time the academic center of the Western world, home to a famous university and library, at which various scholars had gathered, creating a relatively developed research environment for mathematics, astronomy, and medicine. There, Archimedes studied under leading experts, including the disciples of Euclid, and laid the foundations for his future scientific research. He is said to have invented the screw pump (a device for lifting water, which became widely used by the Egyptians) during his time in Alexandria.

The City of Alexandria

Little else is known about this period of study in Alexandria. It is likely that the famous mathematician Euclid was no longer alive, or at least no longer teaching, by this time: although the date and place of his birth are not known, Euclid should have been teaching roughly during the reign of Ptolemy I (ca. 323–285 BCE). There were however at least three notable figures among the fellow students and close friends that Archimedes encountered in Alexandria: Conon of Samos, Dositheus, and Eratosthenes of Cyrene.

Like his Samian predecessor Pythagoras, Conon of Samos was a mathematician and astronomer, who became the closest and most trusted of Archimedes's friends, a friendship that lasted a lifetime. Later he served as court astronomer for Ptolemy III. His work on conic sections formed the basis for the fourth book the famous treatise *Conics* by Apollonius of Perga.

Eratosthenes hailed from Cyrene in North Africa (now Shahhat in Libya). He was about 10 years younger than Archimedes and developed a reputation as a second Plato; he went on to become the director of the Library of Alexandria. Eratosthenes was a man of many accomplishments and interesting personal affects: he was known to be a particular dresser and died at the age of 80 by voluntary starvation in response to the onset of blindness. He wrote a history of ancient drama in ten volumes, earned the nickname *pentathlos* for his intellectual and physical well-roundedness, and created the sieve method in mathematics, which has remained a fruitful technique in number theory through to the present day, both in its original formulation and via various extensions. He measured the circumference of the Earth, obtaining a figure different from the correct one by only 200 km, and deduced that the Atlantic and Indian oceans are connected on the basis of the ebb and flow of their tides, which proved essential to the effort by fifteenth-century Portuguese explorer Vasco da Gama to reach India by water. Finally, Eratosthenes was an influential geographer, introducing the division of the Earth into five climatic zones still in use today, separated by the polar circles and intermediate circles of latitude.

After returning to his native Syracuse, Archimedes maintained a correspondence with Conon and Eratosthenes; to the former he sent his treatises *Quadrature of the Parabola*, *On Spirals*, and *On the Sphere and the Cylinder*, while Eratosthenes received *The Method of Mechanical Theorems* and *The Cattle Problem*. Through these two interlocutors, his work was also communicated to their colleagues in Alexandria, while Eratosthenes and Conon in turn

2 Archimedes: The God of Mathematics

Archimedes Thoughtful (by the Italian painter Domenico Fetti, 1620)

conveyed to Archimedes their own progress. In particular, the fourth century mathematician Pappus attributed to Conon the discovery of the famous Archimedes spiral, which can be seen in the arrangement of the 20 arrondissements of Paris and also made an appearance in the closing ceremony of the 2004 Olympic Games in Athens. Unfortunately, Conon's writings have not survived. These included a treatise in seven volumes entitled *De astrologia* and a correspondence with the mathematician Thrasydaeus dealing with conic sections.

After the death of Conon, Archimedes continued to correspond with Dositheus, a friend or student of Conon who seems to have carried out research primarily into calendrical science and meteorology. Archimedes wrote at the beginning of their correspondence that "Having heard that Conon has died, who was a very dear friend of mine, and that you have been an acquaintance of his and are a student of geometry… I determined to write and send you some geometrical theorems, as I have been accustomed to write to Conon." From various preambles to Archimedes's other works, it is clear that Dositheus also wrote regularly to Archimedes concerning mathematical problems and theorems, but it is not clear what was the specific content of these letters. In any case, it is through these correspondences that the scholarly achievements of Archimedes came to light and were preserved for posterity.

This was by no means uncommon in ancient Greece, since there did not yet exist anything like the modern academic journal, and the publication of books was no easy task at the time. Many scholars announced their results to the world via letters to their friends, the contents of which eventually came to serve as prefaces to their treatises. Apollonius, who lived slightly later than Archimedes, followed the same practice; along with Archimedes and Euclid, he is considered one of the three great mathematicians of the golden age of Alexandria. Like Archimedes, Apollonius studied in Alexandra in his youth and later visited the kingdom of Pergamon, to the north of Miletus in Asia Minor, where there was a library second only to that of Alexandria in size. Apollonius met there a scholar named Eudemus and a figure about whom little is known named Attalus, to whom he sent, respectively, the first three and last five books of *Conics*, securing for them a place in the history of mathematics. This Eudemus however was not the same the more famous historian of science and mathematics of the same name, who came rather from Rhodes and studied with Aristotle.

The Father of Mechanics

Archimedes was related to King Hiero II, the ruler of Syracus at the time, and a friend of his son Gelo II, who later shared the rule with Hiero. There is a famous anecdote related to Hiero and Archimedes which is recounted in the ninth volume of the ten-volume masterpiece *De architectura* by the first-century BCE Roman architect and writer Vitruvius. According to this story, Hiero wanted to repay the gods for the growth of his political prestige and power by building a splendid temple adorned with a crown of pure gold in the shape of a votive wreath. The goldsmith completed this project on schedule and was due to receive his reward, but a rumor emerged that he had kept some part of the gold for himself and replaced it in the crown with a portion of silver. Naturally, Hiero was furious, but he had no way to verify the truth of this rumor, so he asked Archimedes to call upon his wisdom and skill to find a way to test the crown.

At first, Archimedes too could not think of any good solution. In his frustration, he took to the public baths for a break. As he sank into the wooden barrel filled with water, he noticed some of the water flowing over the edge of the barrel and the lightness in his body and suddenly hit upon the key insight. Archimedes realized that although two objects of different material may have

2 Archimedes: The God of Mathematics

Archimedes and the principle of the lever on a Nicaraguan stamp

the same weight, they should displace unequal quantities of water on account of the difference in their volumes. By this principle, it was possible not only to determine whether or not the crown had been made with adulterated gold, but even the amount of the missing gold. Legend has it that Archimedes jumped out of his bath with joy and ran home naked through the streets shouting "Eureka!", which means "I've found it!" in the Dorian dialect. Archimedes was able to expose the misdeeds of the goldsmith, but more importantly he was able to formulate his insight at the theoretical level as the buoyancy principle of hydrostatics: the magnitude of the buoyant force acting on an object submerged in fluid is equal to the weight of the displaced fluid.

This principle appears in the treatise *On Floating Bodies* by Archimedes, which makes no mention however of the anecdote of the crown. Rather it is because of this anecdote that *De architectura* became known to mathematicians; later, following upon the Renaissance, this work also served as a pillar of architecture in the classical period. Sometime around the year 1500, the Italian painter Leonardo da Vinci made his pen sketch of the *Vitruvian Man* based on the requirements of human proportions and the golden ratio presented in the third book of *De archictura*; this became the most famous such drawing in the history of art. In his own time, Vitruvius was probably known by the surname Pollio, later generations opting to refer to him his middle name to avoid confusion with his contemporary, the poet, orator, and historian Gaius Asinius Pollio.

Another anecdote concerning Archimedes is the story of the law of the levers, which appears in the *Mathematical Collection* by Pappus of Alexandria. This law states that a lever is balanced when the distances between two objects, respectively, and its fulcrum are inversely proportional to the weights of the objects. This law provides the foundation for mechanics, and Archimedes is said to have made the bold claim: "Give me a place to stand and with a lever I will move the whole world." A less pithy but perhaps more accurate version of this statement is given by Plutarch, a Greek biographer of the Roman era in his biography of Marcus Claudius Marcellus. According to Plutarch,

Archimedes boasted to King Hiero that with a given force any given weight could be moved and that if there were another earth one could move this one from it. Naturally the king was astonished to hear such a claim and asked Archimedes for a proof by demonstration.

In response, Archimedes chose a cargo ship with three masts from the royal fleet, one which ordinarily required the strenuous effort of a team of men to tow. He installed a set of pulleys, stood alone at a suitable distance holding the rope, and easily pulled the ship. According to the fifth-century Byzantine philosopher Proclus, this ship was built by Hiero as a gift for Ptolemy, the king of Egypt, and required the efforts of almost every Syracusan to launch it by conventional means; Archimedes made it possible for the king to tow it single handedly using a device of his own inventions. Hiero developed after this event a deep appreciation for Archimedes and made his admiration public with the declaration that "From this day forth, we must believe everything that Archimedes says." The author noted with some interest that to this day every giant ship passing through the Panama or Suez Canal does so by means of a pulley car on a track.

The God of Mathematics

Archimedes was not only noble by birth but also aristocratic in spirit, and did not personally attach much significance to his practical inventions, as can be seen from the contents of his recorded works, which consist almost exclusively of problems and results in pure mathematics, his mechanical inventions by contrast coming to us by way of secondhand accounts. Nevertheless, his interest in mechanics exerted a deep influence on his mathematical thinking. Probably his finest mathematical work was *On the Sphere and the Cylinder*, with a preface consisting of his letter to Dositheus. This work starts with six definitions and five assumptions (what we would now call axioms), for example, that among all lines sharing the same two points as extremities, the straight line is the shortest, and similarly that among all surfaces bounded by the same plane curve, the plane has the smallest area. The most famous of the assumptions is known today as the Archimedean property. In the language of modern mathematics, this is the property that given any two positive numbers a and b there exists some natural number n such that $na > b$. From these definitions and axioms, Archimedes goes on to derive some 60 propositions.

For example, Archimedes discovered and proved that the area of a sphere is equal to four times the area of a great circle on its surface and that the volume of a spherical ball is equal to four times the volume of a cone with the great circle of the sphere as its base and the radius of the sphere as its height. It follows from this that the volume of a cylinder with the great circle of the sphere as its base and the diameter of the sphere as its height is three-halves of the volume of the sphere, which is precisely the famous formula for the volume of a spherical ball:

$$V = \frac{4}{3}\pi R^3.$$

This is proved in Proposition 34, a result of which Archimedes was so proud that he requested that the accompanying diagram be inscribed upon his tombstone. Some 700 years later, the same result was obtained by the Chinese mathematicians Zu Chongzhi and his son Zu Geng of the Jin dynasty, using a geometrical construction called the box-lid proposed by the mathematician Liu Hui in the third century.

Another exemplary result is contained in Proposition 14, which states that the surface area of an isosceles cone (excluding its base) is equal to the area of a circle the radius of which is a mean proportional between a generating side of the cone and the radius of the base of the cone, equivalently the product of π, the radius of the base, and a generating side of the cone. In ancient Greece, however, the existence of line segments of certain lengths had become problematic after the discovery by the Pythagorean mathematicians of the irrationality of $\sqrt{2}$, which triggered the first crisis in the history of mathematics. Although this crisis was resolved two centuries later by Eudoxus of Cnidus, who introduced the concept of incommensurability, mathematicians still preferred at the time to avoid invoking the lengths of line segments; for this reason, Archimedes preferred to express his result in terms of the area of a rectangle.

Proceeding the Archimedean principle, Archimedes supplied a rigorous proof of a lemma invoked in Euclid's *Elements* in the course of a proof by a technique known as the method of exhaustion that the difference between the area of a regular polygon circumscribed about a circle and the area of a regular polygon inscribed within it can be made arbitrarily small by requiring a large enough number of sides. This technique was invented by a fifth-century BCE Athenian orator and statesman named Antiphon of Rhamnus in pursuit of a solution to the classical problem of squaring the circle. His idea was to exhaust the area contained within a circle by filling it out with successively larger regular polygons. The method of exhaustion was made rigorous however by Eudoxus, who observed that by subtracting from a given quantity an amount larger than its half and repeating the process as often as necessary, any quantity could be made arbitrarily small.

Archimedes refined the method of exhaustion further still and applied it extensively to solve for the surface areas and volumes of various rotating bodies. For instance, he determined the area of the curvilinear triangle made by the x-axis and the curve $y = x^2$ on the interval [0, 1] by dividing that interval into n equal parts and taking the sum of the rectangular strips so obtained. Unfortunately, his approach to the areas of other such curvilinear triangles required approximation by various special ordinary triangles so that the calculations were not general enough to be extended to the case of generic curvilinear trapezoids.

Measurement of a Circle is a slender work containing only three propositions, all of them concerning the area and the circumference of a circle, but it is not a work to be underestimated. Although Euclid had discussed already many of the properties of circles in his *Elements*, he made no mention of the numerical value of the ratio π or formulas for the calculation of the area and circumference of a circle. Archimedes corrected this deficit in Proposition 1 of this work, which states: the area of any circle is equal to a right-angled triangle in which one of the sides about the right angle is equal to the radius and the other to the circumference of the circle; simply stated, the area is given by the product of the radius and half the circumference. An equivalent statement appears in the ancient Chinese mathematical treatise *Nine Chapters on the Mathematical Art*, which includes the rule that the product of the half the circumference and half the diameter gives the area, which rule reappears in the commentary on this treatise by Liu Hui in 263 CE.

Proposition 3 establishes strong upper and lower bounds for the circumference of a circle with unit diameter, specifically

$$3\frac{10}{71} < \pi < 3\frac{1}{7}.$$

This approximation was obtained from the method of exhaustion by calculating directly the perimeters of the inscribed and circumscribed regular polygons with 96 sides. This marked the first time in history that upper and lower bounds were used to give an approximation for some numerical quantity and provide an estimate of the error. It is worth noting moreover that the fractions on either side of this inequality have the form of truncated continued fractions, so that they are the best possible approximations with denominator 7 or 71. In modern notation, this determines the first two decimal places, $\pi \approx 3.14$, the best approximation obtained by anyone prior to the Common

2 Archimedes: The God of Mathematics   25

Sphere inscribed in a cylinder, symbol of the tomb of Archimedes

Era. The best result up to that point had been the value $\pi \approx 3.1$, used by the ancient Egyptians. Both the ancient Babylonians and the *Nine Chapters on the Mathematical Art* made do with the approximation $\pi \approx 3.0$.

In *On Conoids and Spheroids*, Archimedes studied the areas of ellipses and the volumes of certain solids of revolution. He refined further the method of exhaustion, bringing it very close to the modern methods of integral calculus. In *On Spirals*, he investigated the area enclosed by the first turn of the spiral and the polar axis, as well as the determination of tangent lines of calculus, which touches upon the ideas later developed in differential calculus. The spiral in question is determined by the trajectory of the point moving uniformly along a line rotating uniformly about a fixed point; in modern polar coordinates (invented by Newton), it is determined by the equation $r = a\theta$. The twentieth-century American historian of mathematics E.T. Bell has observed that Archimedes more or less invented the integral calculus some two millennia in advance of Newton and Leibniz and in one of his propositions (referring to *On Spirals*) even anticipated the differential calculus. It is little wonder then that the first-century Roman naturalist Pliny the Elder, the author of *Naturalis Historia*, praised Archimedes as *the god of mathematics*.

Archimedes left behind exactly one work of arithmetic, *The Sand Reckoner*, very possibly his last completed work. This short piece was a witty bit of recreational mathematics, full of imagination and intended for a lay audience, and he dedicated it to the king of Syracuse, Gelo, son of Hiero. This makes it the first work of popular science in the history of the world. Its contents comprise exactly one theorem, equivalent to the modern law of exponents. He also gave estimates for the size of the earth, the moon, and the sun and calculated on the basis of these estimates and the heliocentric model of Aristarchus of Samos the number of grains of sand required to fill up the universe. Of course, he was obliged to make many assumptions, and his calculations differ significantly from the correct values. He also needed to invent a new notation (based on tens of thousands) for the expression of large numbers. His conclusion was that the numbers of grains of sand required to fill the solar system should be about 10^{51} and to fill the entire universe about 10^{63}.

We turn in conclusion to consider the impact of Archimedes's mathematic works on later generations. Although his work was highly original, as, for example, his calculations for the surface area of a sphere and the volume of a spherical ball or his use of the approximation 22/7 for π, its impact in antiquity was very limited, and his work was not developed further; no one, for example, attempted to generalize his formulas for the volumes of solids of revolution, even into the eighth and ninth centuries when his works were translated into Arabic. But after the advent of the Renaissance, such intellectual giants as Filippo Brunelleschi, who designed the dome of the Florence Cathedral, and Leonardo da Vinci became fascinated by Archimedes, to the extent that the former was sometimes referred to as a second Archimedes; at this time, however, they had access only to some handwritten manuscripts. In 1544, seven of Archimedes's mathematical works were finally translated into Latin and published in Basel, and their influence was reflected in the writings of the leading mathematicians and physicists of the time, including Galileo Galilei and Johannes Kepler, and even more substantially in the works of René Descartes and Pierre de Fermat in the seventeenth century. It is to be lamented however that the work with the most potential to exert a groundbreaking influence on the development of mathematics, *The Method of Mechanical Theorems*, was not rediscovered until the early twentieth century.

The Archimedes Palimpsest

It was in 1906 that the Danish philologist and historian Johan Ludvig Heiberg (1854–1928) became aware of the existence in an obscure Greek Orthodox monastery library in Constantinople (now Istanbul) of a mathematical palimpsest that had been partially written over as a prayer book; 2 years later he was able to compile detailed photographs of some 185 pages of the original text, which he prepared for publication in 1915. This text, known now as the *Archimedes Palimpsest*, included two works previously believed to have been lost completely and which had never been translated into either Arabic or Greek, the more important of which was *The Method of Mechanical Theorems*, hereafter referred to as it usually is as *The Method*. *The Method* was one of the treatises that Archimedes had sent to Eratosthenes, in which he explained the process by which he arrived at the conclusions for which he provided only formal proofs in his published works by comparing in his imagination a surface or solid of known area or volume, respectively, with the figure in question. Once he had obtained the correct value, it became straightforward to work out a proof of the claim. The situation is a bit similar to the working methods of modern number theorists, who rely on imagination and computer searches to generate numerical data and empirical observations before attempting their proofs; in the latter case, however, the proofs have rarely been easy to generate.

In *The Method*, Archimedes explained his mechanical method, or method of equilibrium, an alternative to the method of exhaustion, which is better suited to the rigorous proof of already discovered results, but not so helpful in the process of discovery. Archimedes used this method, which showed the influence of atomistic ideas introduced by the philosopher Democritus, to determine the area of volume of certain geometric objects by subdividing them into a large number of thin parallel strips or slices, which he viewed as balanced on one end of a lever at the opposite end of which there is a figure of known area or volume and center of gravity, so that his celebrated law of the lever could be used to determine the unknown quantity. For example, in order to determine the volume of a spherical ball, he placed the ball together with a certain relevant conic solid at one end of the lever and at the other a cylinder of suitable base and height; by arranging everything carefully, he could use the law of the lever and the various known quantities to determine the volume of the ball.

It is clear from this that in addition to the infinitesimal precursors of calculus, Archimedes had in his mathematical quiver a second weapon, consisting of tools from mechanics and physics. We consider two more examples, the first of which is the concept of center of gravity. Borrowing from Newtonian mechanics, this can be easily understood as the single point which stands in, for example, for each of the planets in the course of calculations. In the case of a sphere or a circle, the center of gravity is exactly the same as the geometric center; in the case of a square or parallelogram, it is the intersection of the two diagonals. In the case of a triangle, Archimedes proved in Proposition 1 of *On the Equilibrium of Planes* that the center of gravity lies one-third of the way along any of the lines joining one of the vertices to the midpoint of the opposite side. Turning to the parabola and other conic sections to which Archimedes devoted so much attention, at that time they must have seemed a purely aesthetic or recreational bit of mathematics. But modern science has found that the electrons surrounding the nucleus of an atom, rockets launched into space, and stones shot out from a catapult all follow trajectories given by conic sections.

Patriarch Photius I of Constantinople, a bibliophile who invented the book series and book review

Before tracing the specific history of the *Archimedes Palimpsest*, we discuss the parchment on which it was written. The word *parchment* is taken from its birthplace, the city of Pergamon, which we have mentioned once already. In those times, Pergamon was home to a substantial library and university and had become the academic center for Greek prose and rhetoric. In particular, this city was vying to compete with Alexandria for cultural and academic prestige. So as to hinder this competition, the Ptolemaic dynasty strictly prohibited any export of papyrus to Pergamon, prompting the invention there in the second-century BCE of parchment made of lambskin or calfskin that has been limed, sheared, and softened with a pumice stone. Such parchment paper is smooth on both sides and easy to write on; it is especially well suited to quill pens and can easily be folded into book form. On the whole, it was a better form of paper than papyrus, but also more expensive. For some centuries after its invention, parchment and papyrus were used simultaneously. From the third-century CE until the thirteenth century, parchment paper was the primary form of document until European countries, until it was gradually replaced by Chinese paper starting from the fourteenth century.

We now return to the known history of the palimpsest. In the year 330 CE, Constantine the Great, the first of the Roman emperors to embrace

Ostomachion, from the Archimedes palimpsest

Christianity, built the city of Contstantinople, which became the capital of the Eastern Roman Empire, on the shores of the Bosphorus Strait. He ordered for his new city 50 copies of the Bible and approved a plan for the preservation of classical texts, establishing a reliable scribal tradition. Three centuries later, the Hagia Sophia Cathedral was built, a magnificent building designed by two architects from Asia Minor named Isidore of Miletus and Anthemius of Tralles and widely regarded as a visual and numerical paradigm. Isidore and Anthemius were admirers of Archimedes and had prepared editions of several of his works, which were made more famous by the addition of commentary by their contemporary, the mathematician Eutocius of Ascalon. It is clear then that already by that time Constantinople was home to various writings by and related to Archimedes. These ended up in the collection of Photius I, an ecumenical patriarch of Constantinople in the ninth century who took it upon himself to compile, edit, and publish the various works he had read in the form of series, inventing in the process the modern book review. He also played an important role in the conversion of the Slavs to Christianity, sending two brothers Cyril and Methodius for this purpose, which led eventually to the invention of the Cyrillic alphabet, which is still in use in the languages of more than a dozen countries, including Russia, Ukraine, Belarus, and the Balkans.

In the middle of the ninth century, notational reforms were introduced to the process of transcription replacing capital letters with cursive lowercase script, facilitating an increase in speed and allowing for more text per page. Not long afterward, the Syrian mathematician and astronomer Thâbit ibn Qurra translated the works of Archimedes from Greek into Arabic at the House of Wisdom in Baghdad. In the twelfth century, a translation into Latin was made on the basis of these Arabic editions by the Italian translator Gerard of Cremona, working in Toledo. Not long afterward, Constantinople suffered an unprecedented catastrophe, when, in 1204, it was brutally sacked by the Christian invaders of the Fourth Crusade. Only three codices survived of the works of Archimedes preserved there, referred to, respectively, by the letters A, B, and C. All three of them contained *On the Equilibrium of Planes*, both A and B contained *Quadrature of the Parabola*, both A and C contained *On the Sphere and the Cylinder*, *Measurement of a Circle*, and *On Spirals*, and both B and C contained *On Floating Bodies*; only A included *On Conoids and Spheroids* and *The Sand Reckoner*, and C was the only source for *The Method* and a work called *Ostomachion*. There were of course still other works by Archimedes not included in any of these codices, some of which really have been lost and some, such as a possibly apocryphal collection of geometric problems known as the *Book of Lemmas*, which survived in an Arabic edition.

Today A and B no longer exist except in copies and translations, but they succeeded in the transmission of various treatises and ideas due to Archimedes to the modern era. The remaining codex, which is none other than the palimpsest eventually rediscovered by Heiberg, is therefore not only the only one of them to contain *The Method* and *Ostomachion* (a book which revealed Archimedes's mastery of combinatorics), as well as the Greek version of *On Floating Bodies*, but also the oldest surviving Greek manuscript of any of Archimedes's works. This codex was copied in the tenth century, later erased, and written over sometime in the thirteenth century with a large collection of Orthodox prayers and liturgies, which were preserved in the monasteries as a religious document of the medieval period. The original text remained faintly visible, however, and Heiberg was surprised to find that they were the words of Archimedes. Although the works of Archimedes are not so integrated as Euclid's *Elements*, they are also well-founded and rigorously argued.

Sometime in the 1920s, a French businessman named Marie Lous Sirieix, who had served in Greece, obtained the manuscript while traveling in Turkey and brought it back to Paris. When he moved to the south of France in 1947, he passed it on along with his apartment to his daughter Anne Guerson, who realized its value in 1970 and began attempting to sell it privately. But it was not until 1998 that it was sold for two million dollars at auction on October 29 at Christie's auction house in New York to a wealthy American who preferred to remain anonymous. It now resides at the Walters Art Museum in Baltimore. After years of collaborative effort by a team including experts in the history of science, mathematics, art history, the study of ancient manuscripts, chemistry, digital imaging, and X-ray imaging, this posthumous work finally became available in full to the public. In it, Archimedes proved, among other things, that the ratio of a parabolic segment to the area of an inscribed triangle (as shown in the figure) is 4 : 3, exhibiting yet again the omnipresence of the integer proportions so valued by the Pythagorean school; indeed every proposition in this work is full of magic.

Elegy for a Hero

In the year 212 BCE, the emperor Qin Shi Huang of China ordered the burning of books and burying of scholars at Xianyang, wherein more than 460 Confucian scholars were buried alive. That same year, Archimedes of Syracuse also came to the end of his life. His death is part of the story of the Punic Wars, three wars which took place between Carthage and the Roman Empire between 264 and 146 BCE due to conflicts of commercial, transportation, and colonial interests (their name derives from the Latin name *Poenus* for the Phonecians as applied to the citizens of Carthage, who were of Phoenician ethnicity). The bloodiest of these was the Second Punic War, lasting from 218 to 201 BCE, which had an impact similar to that of the Second World War in the twentieth century. The Carthaginians had the upper hand at one point, especially under the leadership of their young commander Hannibal, who gained complete control over the sea and whose army crossed the Pyrenees and the Alps by land into the Apennine hinterland but finally succumbed to Roman raids on the Carthaginian mainland and were obliged to call back their army for reinforcements.

Since Syracuse was allied at that time with Carthage and lay in the sea passage bringing the Roman warships toward Carthage, it inevitably became a target for Roman conquest. In 214 BCE, the famous Roman general Marcellus led a large army to besiege the city. A record of the siege appears in *The Histories*, by the second-century BCE Greek historian and statesman Polybius, the earliest of many books detailing the history of this war. Polybius states that Marcellus staged his attack by sea, and the Syracusans relied in their defense upon ingenious instruments of war devised by Archimedes, including cranes capable of lifting whole boats out of the water and dropping them back into it with force. Marcellus advanced with eight five-decked ships, chained together in twos, but before they came close enough to the Syracusans to do damage, they were met with a storm of massive stones thrown at them by machinery and forced to retreat after suffering many casualties.

Another legend is recorded in the account of Lucian of Samosata, a second century rhetorician and satirist, who claimed that Archimedes had used a giant mirror to focus such light onto the enemy ships as to light them on fire.

Engraving of Archimedes on the Fields Medal

This may be an exaggeration but suggests that Archimedes was already aware of the nature of parabolic mirrors and their ability to focus light. In response, the Romans took to attacking at night, but Archimedes was prepared for this as well and prepared a type of catapult nicknamed the Scorpion, suited especially to use at close quarters, leading to another large Roman loss. In the end, Marcellus simply abandoned the idea of a direct attack and settled in for a long siege. The city of Syracuse finally succumbed to food deprivation and was quietly captured by the Romans on a festival night in 212 BCE; during this defeat, Archimedes died an honorable but tragic death.

The first account of this comes from Livy, a historian of the Common Era and author of a monumental history of Rome entitled *Ab Urbe Condita* (meaning *From the Founding of the City*), who wrote: "Many brutalities were committed in hot blood and the greed of gain, and it is on record that Archimedes, while intent upon figures which he had traced in the dust, and regardless of the hideous uproar of an army let loose to ravage and despoil a captured city, was killed by a soldier who did not know who he was." Tzetzes adds to this a further detail that Archimedes failed to notice who it was that was approaching him and called out, "Stand away, fellow, from my diagram," leading to his death, whereas Plutarch has it that Archimedes asked the soldier to stay his hand long enough for him to work out the solution to the problem he was working on; intriguingly, this is perhaps the only allusion to Archimedes among the ancient accounts to touch upon pure mathematics.

It is further related that after the death of Archimedes, Marcellus was so grieved that he dealt seriously with the soldier who killed him and sought for the relatives of the scientist with compassion and to honor them with signal favors. He also honored the dying man's wish that his tombstone be engraved with the sphere and the cylinder associated with the mathematical work of which he was most proud. It is of note that Plutarch recorded his version of these events in a biography of Marcellus rather than of Archimedes, believing perhaps that the general was the more important historical figure of the two; in fact, this general is mainly known to us today if at all for his association with the mathematician. More than a century later, the Roman politician Marcus Tullius Cicero, in his capacity as the tax collector at Sicily, made a pilgrimage to visit the tomb; although he could not find anybody willing to show him the way, he was able to find it on his own and observed the pattern of the sphere and the cylinder was still clearly visible. I myself do not know whether the tradition of having figures or formulas upon a tombstone originated with Archimedes. In any case, the tomb of Archimedes was later buried by time.

The English philosopher Alfred North Whitehead famously remarked that "The safest general characterization of the European philosophical tradition is that it consists of a series of footnotes to Plato," to which there has been added the clever gloss, "The safest general characterization of the European scientific tradition is that it consists of a series of footnotes to Archimedes." Today Archimedes is recognized as easily the greatest mathematician and scientist of the ancient world. According to E.T. Bell has observed that "Any list of the three 'greatest' mathematicians of all history would include the name of Archimedes. The other two usually associated with him are Newton and Gauss." But if we compare their various achievements to those of their contemporaries, Archimedes probably occupies the first position. In 1979, the poet Odysseus Elytis, like Archimedes a Greek, who was born in Crete, received the Nobel Prize for Literature. In a long poem entitled *Heroic and Elegiac Song for the Lost Second Lieutenant of the Albanian Campaign*, he wrote: "How the faint smoke of dreams rises …/This moment casts aside another moment/So the eternal sun has left the world".

Summer 2013, Hangzhou, Caiyunju

3

The World of Omar Khayyam

Isfahan is half the world.
—*Persian proverb*

The World of the Body

In order to understand the trajectory of the life of the Persian mathematician and poet Omar Khayyam, it is necessary to start from the historical name of his hometown, Khorasan. This name in Persian means "land from which the sun arrives," indicating the east. Today, Razavi Khorasan is a province in northeastern Iran; its capital Mashhad is a site of pilgrimage for Shia Muslims, and the region is famous for handwoven carpets with exquisite patterns. But in the past, the name Khorasan extended across a much wider area, including the modern province vast tracts of southern Turkmenistan and northern Afghanistan, stretching from the Caspian Sea to the Amu Darya River in the north, from the edge of the deserts of central Iran to the Hindu Kush mountains of Afghanistan in the south. Some Arab geographers have even contended that this region reached all the way to the borders of India.

Incidentally, the Amu Darya is the largest river in Central Asia, meandering through Afghanistan, Tajikistan, Uzbekistan, and Iran before it finally empties into the Aral Sea. According to legend, the ninth century Arab mathematician al-Khwarizmi, who gave us the word *algebra*, was born in the torrid

Statue of Omar Khayyam in Nishapur

ancient city of Khiva, downstream of this river in what is now Uzbekistan. The mountains of the Hindu Kush were historically a center for Buddhism, through which the Chinese monk Xuanzang passed through in his search for Buddhist scriptures. He referred to them as the Snowy Mountains in his travelogue *Great Tang Records on the Western Regions*. The footprints of Omar Khayyam however exceeded even the geographical scope of Khorasan. He made his way to Samarkand, the central city of Uzbekistan to the north, to Isfahan on the Iranian plateau in the south, and even to Mecca, the western tip of the Arabian Peninsula.

As a mathematician, Khayyam lived and worked in so many countries (four, according to modern administrative divisions, excluding the pilgrimage site of Saudi Arabia) that only Pythagoras of Greece, who is said to have spent time in places including Greece, Lebanon, Egypt, Iraq, and Italy, could surpass him. Considering also the poets of the ancient world, although their careers required wide travels across the known world, none of them seem to have been so fortunate as him. Perhaps it is for this reason that Homer had the protagonist of his epic *The Odyssey* suffer through 10 years lost at sea before returning to his hometown, and Dante sent himself literally through hell and heaven in his *Divine Comedy*. It is most likely that Khayyam was able to travel around the world so freely because he was born into a family of craftsmen and benefited from the wide scope of Islam in the region.

Omar Khayyam was born on May 18, 1048, in Nishapur along the ancient Silk Road, today a small city of several hundred thousand about 70 km from the larger capital Mashhad and famous in particular for its legacy of ceramic artistry. He received education first in his hometown and later in Balkh, a town in northern Afghanistan about 300 km northwest of Kabul and thousands away from his hometown. The name *Khayyam* means *tentmaker* in Arabic, suggesting that his father was an itinerant craftsman, moving with his family from one city to another. The times too were turbulent: as Khayyam wrote in the preface to his *Treatise on Algebra*, "I cannot concentrate on learning algebra, the turmoil of the present situation hinders me." Nevertheless, he wrote quite a lot including his valuable *Difficulties of Arithmetic* and a pamphlet on music.

Sometime around the year 1070, in his early 20s, Khayyam left his hometown and traveled north to Samarkand, one of the oldest cities in Central Asia. Samarkand had earlier been conquered by Alexander the Great, and at that time, it was under Turkish control. This was prior to the birth of both Genghis Khan, the fearsome leader of the epoch, and the Italian traveler Marco Polo, both of whom later stepped foot on that land, coming from different directions and by different ways. Khayyam had been invited by a renowned local scholar with political status and influence, under whose patronage he engaged in mathematical research with an untroubled mind and made important discoveries in algebra, including a geometric solution to cubic equations. This was considered the most profound and innovative mathematics of the period. On the basis of these achievements, Khayyam completed the work that brought him fame, entitled *Treatise on Demonstration of Problems of Algebra* but frequently referred to as the *Treatise on Algebra*.

Not long afterward, Khayyam was invited by the third sultan Malik-Shah I of the Seljuk Empire to travel west to Isfahan to carry out astronomical observations there and implement calendar reform. He was moreover directed to oversee the construction of an observatory in the city. The Seljuks were originally the ruling family of the Oghuz alliance of Turkic tribes living in Central Asia and the Mongolian grasslands. One of these had settled in the lower reaches of the Syr Darya, the longest river of Central Asia, in modern day Kazakhstan near the Aral Sea, and joined the Sunni Islamic sect. In the eleventh century, they suddenly uprooted and traveled first south, and then west, becoming in the process a large empire controlling the territories from the Amu Darya to the Persian Gulf and from the Indus River to the Mediterranean Sea. There can be no doubt that the Mongolian expeditions a century later took this as their inspiration; Mongolians and the Turks shared a common ancestry, although by contrast only some of the Mongols converted to Islam.

The Seljuks did not have their own such cultural traditions—they accepted the language of the Persian scribes within their jurisdiction. Persian literature achieved a wide circulation, and Persian scholars and artists were widely respected; much the same had happened during the Macedonian conquest of Greece. It was because of this that Khayyam had the opportunity to move to the capital. So we must turn now to the city of Isfahan. Today it is the second largest city in Iran, behind the capital, Tehran, with a population of more than a million, famous for its magnificent mosques, spacious squares, canals, tree-lined avenues, and bridges (such a scene was vaguely in evidence when I arrived for a visit in late summer in 2004). Centuries after the time of the Seljuk Empire, King Abbas I of the Persian Empire also made Isfahan his capital, earning it a reputation as the most beautiful and moving city of the seventeenth century. There is a Persian proverb that survives to this day: "Isfahan is half the world."

Malik-Shah I was the most famous sultan of the Seljuk Empire. In 1072, he inherited the throne at the age of 17 with the full support of the aging *vizier* of the empire Nizam al-Mulk. During his reign, Malik-Shah inherited the business of his father, conquered the rulers of Upper Mesopotamia and Azerbaijan, annexed the territories of Syria and Palestine, and gained control of Mecca, Medina, Yemen, and the Persian Gulf region. It is said that one of his armies arrived at Nicaea on the far side of Constantinople and took it over. The Byzantine Empire in response sent envoys to the west to ask for aid, which led some years later to the First Crusade. During this time, the people of the country lived and worked in peace and contentment. The sultan himself showed great interest in literature, art, and science. He invited many scholars to visit and treated them well, promoted education, and developed various scientific and cultural undertakings.

According to the historians, Isfahan during the reign of Malik-Shah was famous for its golden mosques, the poems of Omar Khayyam, and calendrical reform; these latter two were both directly related to the subject at hand.

3 The World of Omar Khayyam

Footprints from the life of Omar Khayyam

The Mausoleum of Omar Khayyam in Nishapur

Without a doubt, this was the most peaceful period in the life of Khayyam, and he served as director of the observatory for 18 years. Unfortunately, in 1092, the brother of Malik-Shah, at that time governor of Khorasan, launched a rebellion and sent assassins to carry out the murder of al-Mulk. The sultan died suddenly as well in Baghdad, and the Seljuk Empire went into a steep decline. When the second wife of Malik-Shah took the reigns of political power, she looked unkindly upon Khayyam and withdrew funding for the observatory. His calendrical reform also became difficult to carry out, and his research ground to a halt. Nevertheless, Khayyam stayed on, hoping to wait out the changes to his political situation and improve his lot through his own powers of persuasion.

Around the year 1096, the third son Ahmad Sanjar of Malik-Shah became the last sultan of the Seljuk Empire. By this time its territory had already diminished, and he served more as the monarch of Khorasan. As an adult, Sanjar managed to conquer the lands between the Amu Darya and the Syr Darya and reach the Indian border, but he was ultimately defeated at Samarkand, and in 1118 he was compelled to move his capital to Merv, an ancient city in Central Asia, the ruins of which are located near Mary, a regional capital in modern Turkmenistan. Khayyam accompanied him, and while they were there, he and disciples wrote together a book entitled *On the Deception of Knowing the Two Quantities of Gold and Silver in a Compound Made of the Two*, which explored a problem with famous pedigree dating back to Archimedes concerning the use of mathematical methods to determine the composition of alloys from their specific gravities.

In his later years, Khayyam returned alone to his hometown Nishapur, in fact not far from Merv, where he attracted to him several disciples and occasionally provided predictions for the court. He never married and left behind neither children nor any inheritance. After his death, his students buried him under peach and pear trees in the suburbs of the city. In the middle of the nineteenth century, his quatrains were translated into English, and his reputation as a poet spread across the globe. To date, the *Rubaiyat of Omar Khayyam* has been published in more than a hundred editions in dozens of countries. In 1934, various countries contributed funds to the erection of a tall mausoleum in his memory in his hometown. The Mausoleum of Omar Khayyam is a complex geometric structure, surrounded by eight pointed prisms inlaid with beautiful patterns in the tradition of Islamic art.

The World of the Mind

The early mathematical works of Omar Khayyam have been lost to history, and only the cover and a few fragments of his *Difficulties of Arithmetic* are known today, housed at Leiden University in the Netherlands. On the other hand, his *Treatise on Algebra*, certainly one his most important works, survived. This book was translated from Arabic into French by Franze Woepcke in 1851, under the title *L'algèbre d'Omar Alkhayyami*. Although it had been missed during the era of translations in the twelfth century, still this work appeared some 8 years earlier than the English translation of the *Rubaiyat*. In 1931, in celebration of the 800th anniversary of his birth, an English translation of the *Algebra of Omar Khayyam* prepared by Daoud Suleiman Kasir was published by the *Teachers College Press* in New York. Our knowledge of the mathematics of Omar Khayyam today is based mainly on translations of this work.

At this start of the *Treatise on Algebra*, Khayyam recalls some results from his earlier work, *Difficulties of Arithmetic*: "From the Indians one has methods for obtaining square and cube roots ... We have written a treatise on the proof of the validity of those methods and that they satisfy the conditions. In addition we have increased their types, namely in the form of the determination of the fourth, fifth, sixth roots up to any desired degree. No one preceded us in this and those proofs are purely arithmetic, founded on the arithmetic of the *Elements*." In particular, the earlier book made mention of here is most likely the *Difficulties of Arithmetic*, while the *Elements* is the famous mathematical work of Euclid, which had been translated into Arabic in the ninth century; it was translated much later and only partially into Chinese by the Italian missionary Matteo Ricci and the Ming Dynasty scholar Xu Guangqi in the seventeenth century.

This Indian algorithm with which Khayyam was familiar came primarily from two early Arabic works: the *Principles of Hindu Reckoning* and *Kibab al-Fusul fi al-Hisub al-Hindi*. But because he spent his early years on the ancient Silk Road connecting Central Asia and China, it is likely that he was also inspired and influenced by Chinese mathematics. The *Nine Chapters on the Mathematical Art*, which was published by the first-century BCE, gives a complete set of rules for the determination of square and cube roots. Among the surviving Arabic literature, the earliest systematic treatment of higher powers of natural numbers and the extraction of roots appears in a treatise on arithmetic by means of dust board compiled in the thirteenth century by the polymath Nasir al-Din al-Tusi; al-Tusi does not indicate the origins of his methods, but since he was familiar with the work of Omar Khayyam, historians of

Ancient buildings in Isfahan (photograph by the author)

mathematics speculate that they most likely appeared in the *Difficulties of Arithmetic*, but this cannot be confirmed because that work has been lost.

The greatest mathematical achievement of Omar Khayyam was the use of conic sections to solve cubic equations; indeed, this is the most remarkable work by any medieval Arab mathematician. The conic sections comprise ellipses (among which circles are included), hyperbolas, and parabolas, in short the familiar curves that we encounter in middle school and which can be obtained as the intersection of a cone and a plane. The problem of solving cubic equations can be traced back to the famous problem from ancient Greece of doubling a cube, specifically to find a cube whose volume is twice that of a given cube. Written as an algebraic equation, it is necessary to solve $x^3 = 2a^3$. In the fourth-century BCE, the Greek mathematician Menaechmus

3 The World of Omar Khayyam 43

of the Platonic school discovered the conic sections and transformed this problem into one of determining either the intersection of two parabolas or the intersection of a parabola and a hyperbola. Such problems attracted intense interest from Islamic mathematicians, and it was to the credit of Khayyam that he considered every form of cubic equation and solved them separately in turn.

In detail, Khayyam divided the forms of cubic equation into 14 categories, including one category of equations with no linear or quadratic terms, 3 categories with either no linear terms or no quadratic terms, and 7 categories that did not exclude any of the possible terms; the roots of the equations in the various cases can then be determined from the intersection of two conic sections. Take as an example the equation $x^3 + ax = b$, which can be rewritten as $x^3 + c^2x = c^2h$. According to Khayyam, the solution of this equation is exactly the abscissa x of the point C of intersection of the parabola $x^2 = cy$ and the semicircle $y^2 = x(h - x)$ (as shown in the figure), because eliminating y from these two equations gives the previous one. However, Khayyam used prose in place of equations in describing this solution, which makes it very difficult for readers to follow. This is one reason why Arabic mathematics, like ancient Chinese mathematics, later faced obstacles to further development.

Khayyam also endeavored to find arithmetical (or algebraic) solutions to cubic equations, but without success. Nevertheless, he predicted in his *Treatise on Algebra* that future generations of mathematicians might find solutions at least for those equations lacking constant terms, linear terms, or quadratic terms, a prediction realized some five centuries later when Italian mathematics worked out algebraic solutions for cubic and quartic equations. As for polynomial equations of degree five or higher, it was proved in the nineteenth century by the Norwegian mathematician Niels Henrik Abel that no general solution in radicals exists. It is worth mentioning that the history of polynomial equations in Europe was a fractious one: priority disputes over the solution of cubic and quartic mathematicians brought several Italian mathematicians into conflict with one another, and the work of Abel was not appreciated by his contemporaries until after his death.

In the field of geometry, Khayyam also made two contributions. One was to provide new insights on the issue of ratio and proportion, the other a critical discussion and demonstration of the parallel postulate. Ever since Euclid's *Elements* had been introduced to Islamic scholars, the fifth postulate attracted the attention of mathematicians. This postulate consists of the following geometric axiom: "If a line segment intersects two straight lines forming two interior angles on the same side with angle sum less than two right angles, then the two lines, if extended indefinitely, must meet on that side." This

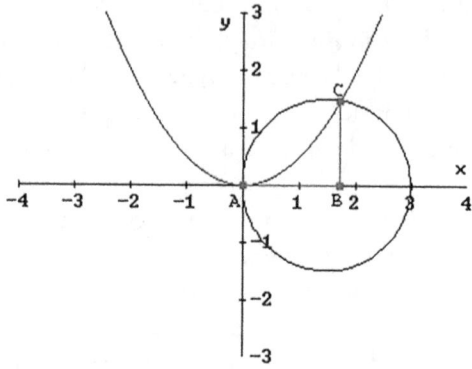

Geometrical solution of the cubic equation

postulate is more complicated than the other four introduced by Euclid in both its structure and content, and not so immediately obvious, leading people naturally to a desire to either prove it from the others or replace it with something simpler. It should be pointed out here that a simpler version of the fifth postulate was indeed provided by the eighteenth-century Scottish mathematician: "In a plane, given a line and a point not on it, exactly one line parallel to the given line can be drawn through the point." Even this version however is not self-evident.

In the year 1077, while in Isfahan, Khayyam wrote a new book entitled *Commentary on the Difficulties Concerning the Postulates of Euclid's Elements*, in which he attempted to prove the fifth postulate from the prior four. He considered a quadrilateral *ABCD* (as shown in the figure), assuming that the angles at *A* and *B* are both right angles, and the lengths of the line segments *CA* and *DB* are equal; by symmetry the angles at *C* and *D* are equal. He realized that the fifth postulate would follow immediately from a demonstration that also the angles at *C* and *D* are right angles. Therefore, he took as assumptions that the two angles were either obtuse or acute and tried to obtain a contradiction in each of these two cases. It is intriguing that this approach to the problem is closely related to the development of non-Euclidean geometry in the nineteenth century. In fact, non-Euclidean geometry, one of the most important discoveries in modern mathematics, follows directly from the assumption that one or the other of these two hypotheses does in fact hold.

Unfortunately, Khayyam did not realize this, and his argument was predestined to failure. Instead, he proved that the parallel postulate could be replaced

Quadrilateral used in the attempted proof of the parallel postulate

as an assumption by the following one: if two straight lines are drawing closer to one another in one direction, then they must eventually intersect in the same direction. Here we mention in passing that one of the foundation figures of non-Euclidean geometry, the Russian mathematician Nikolai Lobachevsky, lived in Kazan, likewise far from the centers of European civilization. Kazan is the capital of the Tatar Autonomous Republic, a gathering place for ethnic minorities. Like Isfahan, it is located at about the meridian 50° east of Greenwich, but Kazan is north of the Caspian Sea, while Isfahan is to its south. Although Khayyam failed to prove the parallel postulate, his approach to this issue influenced later Western mathematicians by way of al-Tusi, most notably among them John Wallis, the direct mathematical predecessor to Isaac Newton.

In addition to his mathematical research, Khayyam also led his group of astronomers in Isfahan to compile an astronomical table, named the *Astronomical Handbook with Tables for Malik-Shah* in honor of his patron. Only a small portion of this table has survived to the present day, including a table of ecliptic coordinates and the hundred brightest stars. More important than this was his work on calendrical reform. Since the first-century BCE, the Persians had used the Zoroastrian solar calendar (this religion goes back to the seventh-century BCE), dividing the year into 12 months and 365 days. After the Arab conquest, they were compelled to adopt the Hijri calendar, which is like the Chinese lunar calendar: long months have 30 days, short months have 29, and the whole year has 354. But whereas the Chinese calendar incorporates leap months to make for consistent winter and summer seasons, the

Hijri calendar is religious in nature and incorporates 11 leap days every 30 years, to the detriment of agriculture, summer falling sometimes in the sixth month and sometimes in the first.

When Malik-Shah was in power, the Persians had reintroduced the solar calendar, and the sultan insisted on further calendrical reform as part of the duties of his observatory director. Khayyam suggested that on the basis of 365 days in an ordinary year, there should be 8 leap days per 33 years, establishing an average year length of 365 days and 8/33rds of a day, less than 20 s off from the actual tropical year, that is, the time it takes for one rotation about the Sun. In particular, the difference comes to 1 day every 4460 days, less than the modern international standard. The Gregorian calendar, which includes 97 leap days every 400 years, suffers by a difference of 1 day in every 3333 days. This calendar was established by Pope Gregory XIII in 1582, but it was not implemented by countries outside the Catholic sphere, including the United Kingdom, the United States, Russia, and China, until the eighteenth, nineteenth, or even the twentieth century. Considering this issue mathematically, it is interesting to look at the asymptotic fractions obtained from expanding the decimal part of the tropical year, which are

$$\frac{1}{4}, \frac{7}{29}, \frac{8}{33}, \frac{31}{128}, \frac{132}{545}, \cdots$$

The first of these corresponds to a leap day every 4 years, as in the Julian calendar promulgated by the Roman emperor Julius Caesar when he was still a consul. This gives an error of 1 day every 128 years. The Khayyam calendar corresponds to the third fraction, 8/33. In light of this, it is mathematically the best possible calendrical formulation within the limitations of a 128-year range. He took March 16, 1079, to be its starting point and called it the *Malik-Shah Calendar*. Unfortunately, this work was abandoned midway through upon the death of his patron. At that time, the solar calendar in use in countries around the world was already in error by more than 10 days. Khayyam expressed his disappointment in a poetic lamentation, No 57 according to the numbering of the *Rubaiyat*:

> Ah, but my Computations, People say,
>   Reduced the Year to better reckoning?—Nay
> 'Twas only striking from the Calendar
> Unborn To-morrow, and dead Yesterday.[1]

---

[1] Tr. Edward FitzGerald.

The World of the Spirit

If Omar Khayyam had only been a mathematician and astronomer (in fact he was also said to have been proficient in medicine and to have served as physicians to the sultan), then perhaps he would not have lived alone the entirety of his life, although his intellectual successors such as Descartes, Pascal, Spinoza, Newton, and Leibniz also never married. But these intellectual giants of the west, in addition to their scientific research, were all dedicated in spirit to religion or philosophy. Khayyam on the other hand recorded his inner life beyond the confines of the kingdom of science in the form of poetry. But his works were likely put to the side after initial exhibition because they did not agree with the times; or perhaps they were ignored as the incidental pastime of a mathematician and astronomer. It is agreed upon however among later scholars that the poetry of Khayyam was not much beloved by orthodox Muslims, as he was not constrained in it by the Islamic view of God's creation of the world. There is less agreement among scholars as to the actual number of poems he composed.

In order to discuss the poetry of Omar Khayyam, it is necessary to understand somewhat the Persian literary tradition. In the year 651 CE, Muslim conquests brought an end to the Sassanid Empire, the last empire of ancient Iran, and placed Persia under the jurisdiction of the theocratic caliphate. Islam replaced Zoroastrianism, and Arabic took the place of Persian as the official language. Among the Persian people, however, a new language emerged, which became Modern Persian, an offshoot of Pahlavi, or Middle Persian. In the course of its evolution, this language came to be written in the Arabic alphabet and saw the introduction into it of Arabic vocabulary. Persian literature, the literature of Modern Persian, emerged in Khorasan, Khayyam's hometown. Subsequently, notable Persian poets and writers appeared one after another, on the eastern coast of the Mediterranean, in Central Asia, the Caucasus, Afghanistan, and North India.

Moreover, after several centuries of Arabic occupation, a new Persian empire, the Samanid dynasty, appeared far away from the Arabian Peninsula; its territory included Khorasan and the lands between the rivers. Prior to the arrival of the Seljuks, these lands saw nearly two centuries of free development and industrial and commercial prosperity. The city of Samarkand became a center for learning, poetry, and art; another was Balkh in northern Afghanistan—exactly the two destinations at which Khayyam had spent time in his youth. In the middle of the ninth century, Rudaki, regarded as the father of Persian poetry, was born on the outskirts of Samarkand. He traveled widely while he was young, and although later in life he suffered from poverty and blindness, still he lived to the age of 90 and established the style of poetry known as the Khorasan style.

Six years before the death of Rudaki, another important poet was born in Khorasan. This was Ferdowsi, considered by the Persian people to be their greatest poet. His most important work is the narrative poem *Book of Kings*, completed in the year 1010, which relates the history of the heroes and kings of Persia from mythical times through to the Sassanid dynasty. For a millennium since, generations of Persian speakers have recited or listened to this collection of poems. It has the characteristic features of Khorasan poetry: simple words, unfussy descriptions of characters and environments without much use of foreshadowing, and only sparse use of Arabic vocabulary. Although some Western scholars criticized this vast poem as containing monotonous rhythms and boring, repetitive content, these remarks misunderstand the perspective of modern Iranian people, for whom the work is as accessible as is the King James Bible for English-speaking Christians.

Chinese edition of the *Rubaiyat*, translated by Guo Moruo, 1924

Khayyam was born more than 20 years after the death of Ferdowsi, but by this time his hometown had already come under the rule of the Seljuk dynasty. Probably he took to poetry already in Nishapur, but otherwise he must have been inspired to it in the literary capitals of Balkh and Samarkand. Since Khayyam is not mentioned in his capacity as a poet until a half century after his death, we have no way of knowing how he was regarded as a writer during his lifetime. He wrote in quatrains, or four-line verses, a form pioneered by Rudaki. The ends of the first, second, and fourth lines are required to rhyme, similarly to Chinese quatrains. Although there were no strict requirements on the number of words in each line, Khayyam adhered to the principle of saying always enough to astonish. In No. 71 of the *Rubaiyat*, he writes:

> The Moving Finger writes; and, having writ,
>   Moves on: nor all your Piety nor Wit
>   Shall lure it back to cancel half a Line,
>   Nor all your Tears wash out a Word of it.[2]

In 1859, the year Darwin published *On the Origin of Species*, the Englishman Edward FitzGerald compiled 101 of Omar Khayyam's poems into a simple booklet that he called the *Rubaiyat of Omar Khayyam*, the word *rubaiyat* meaning quatrains in Arabic, which he published anonymously. He was already 50 at the time, and unknown to the literary world. In fact he had previously attempted a Latin translation, before settling upon his native tongue. In his early years, FitzGerald had studied at Trinity College, the most prestigious college at Cambridge, and had formed a lifelong friendship with William Thackeray, author of the novel *Vanity Fair*. Upon graduation, he lived life as a country squire and enjoyed the company of such great writers as Alfred, Lord Tennyson, and Thomas Carlyle, although he himself lacked confidence in his own writing. It was only in middle age that he began to learn Persian and took an interest in the East. In his translation of the *Rubaiyat*, FitzGerald was free with his language and did not adhere strictly to the original text, frequently substituting metaphors of his own invention to convey the essence of the poetic thought.

Starting in the second year of its existence, British literary figures began to sing the praises of this translation. The poet and critic Algernon Charles Swinburne wrote that FitzGerald had given Omar Khayyam a permanent place among the greatest English poets, and G.K. Chesterton observed the romantic and classical character of the unparalleled collection, in both its ethereal melody and lasting impression. Other critics have felt that the

---

[2] Tr. Edward FitzGerald.

Chinese edition of the *Rubaiyat*, translated by Tianxin Cai, 2024

translation consists more of English poems with Persian images, which is an exaggeration. The *Encyclopedia Britannica*, for example, lists FitzGerald as a writer rather than a translator. But from his other literary efforts, it is apparent that in fact FitzGerald was rather mediocre as a writer and likely would not merit an encyclopedia entry if not for his translation work.

A Chinese edition of the *Rubaiyat* appeared in 1924, translated and published by Guo Moruo on the basis of the English version. Since that time, more than a dozen Chinese poets and scholars and even one MIT physicist have attempted translations, either from English or from Persian. Guo Moruo compared Khayyam to a Li Bai of Persia (because they were both addicted to alcohol). Intriguingly, it was Guo Moruo who first proved, nearly half a century later, that Li Bai was born in Suyab in Central Asia, near the city of Tokmok on the west coast of Issyk-Kul Lake in modern Kyrgyzstan. Perhaps he wanted to make Li Bai and Khayyam into neighbors. During the Cultural Revolution, Guo Moruo's *Li Bai and Du Fu* (1971) was one of the few treatises available to Chinese public intellectuals. Here I would like to insert a poem (No. 35 of the Rubaiyat) by Khayyam on the subject of wine:

> Then to this earthen Bowl did I adjourn
>   My Lip the secret Well of Life to learn:
>   And Lip to Lip it murmur'd—"While you live,
>   Drink!—for once dead you never shall return."[3]

As the ancients said, benevolence sees benevolence, and wisdom sees wisdom. The Argentine author Jorge Luis Borges wrote of the FitzGerald translation of *The Rubaiyat* that it begins "with images of morning, the rose, and the nightingale" and ends "with those of the night and the tomb."[4] Khayyam like Borges was a contemplative person, struck with an inability to tear himself away from such problems as the ultimate nature of heaven and earth, the brevity and impermanence of life, the relationship between man and god. He was skeptical as to the existence of an afterlife, a heaven and a hell, laughed at the self-righteous certainty of religion and the pedantry of scholars, and lamented human fragility and the harshness of the social environment. He could not find any satisfactory answers to these conquests and turned instead to the worldly enjoyment of sensual pleasures; but he could never avoid them entirely.

This theme, of enjoying what life has to offer while still there is time, was at one time one of the great traditions of European literature according to

---

[3] Tr. Edward FitzGerald.
[4] *The Enigma of Edward FitzGerald*, from *A Personal Anthology* by Jorge Luis Borges.

T.S. Eliot. Its connotation of course includes a negative attitude toward life in the general sense, but also a positive philosophy of life in its particulars. In fact, the topics of wine and beauty appear more often in the poems of Omar Khayyam than they do in those of the bohemian Li Bai, notwithstanding the fact that alcohol is prohibited by the Islamic faith. This is perhaps one reason his poems were denounced as colorful by contemporary scholars—a poisonous snake and a devourer of doctrine. In 2014, when the Iranian mathematician Maryam Mirzakhani became the first woman to win the Fields Medal, she sparked controversy among her compatriots for refusing to wear the headscarf, and the media mostly distributed images from her time in Iran. Unfortunately, Mirzakhani died 3 years later of breast cancer; the next year, however, the Fields Medal went to another Iranian mathematician, Caucher Birkar.

Iranian mathematician Maryam Mirzakhani, winner of the Fields Medal

So it is that in the eyes of the devout, the poems of Omar Khayyam are a nonsensical absurdity. Under pressure from religious leaders, he traveled long distances in his later years to make the pilgrimage to Mecca, the holiest site of Islam, but he himself always sailed against the current and wrote his poems in order to explore the mysteries of life and the value of existence, extracted from inanimate objects (No. 29 of the Rubaiyat):

> Into this Universe, and why not knowing,
>   Nor whence, like Water willy-nilly flowing:
> And out of it, as Wind along the Waste,
>   I know not whither, willy-nilly blowing.[5]

---

[5] Tr. Edward FitzGerald (No. 29).

At the beginning of the last century, T.S. Eliot, then a 14-year-old boy from St. Louis in the United States, stumbled across the FitzGerald translation of the *Rubaiyat* and immediately became fascinated with it. This rare and great poet of the twentieth century later recalled the simply beautiful scene full of bright, sweet, and painful colors into which he entered reading this glorious poetry and credited this experience with his determination to be a poet. In the martial arts novel *The Heaven Sword and Dragon Saber* by Jin Yong, one of the female characters, Yin Li, sings again and again a little song, "It comes like flowing water and passes like the wind; I don't know where it came from, nor where it ends," which was borrowed from the *Rubaiyat* with two characters added to create the flavor of an ancient Chinese poem. At the end of this novel, another of the female characters, Xiao Zhao, was sent to Persia.

Hangzhou, December 2006–February 2015

4

Qin Jiushao, Daogu Bridge, and the *Mathematical Treatise in Nine Sections*

> *What is truth? said jesting Pilate, and would not stay for an answer.*
> —*Francis Bacon*

New Daogu Bridge

In the city of Hangzhou, not far from Baoshi Mountain on the north bank of West Lake, there is a small road called Xixi Road (*Xixi* means West Brook), and toward the eastern stretch of this road, between the Yuquan and Xixi campuses of Zhejiang University, but closer to the Xixi campus, there is a stone bridge known as Daogu Bridge. It was first built during the Jiaxi period of the Southern Song dynasty (1237–1241); at that time, it was called Xixi Bridge. In the *Lin'An Chronicles*, written in the early years of the Southern Song Xianchun period, there appear about it the following remarks: "The Xixi Bridge, east of the examination courtyard of Lin'an Prefecture (Hangzhou) was built by Daogu during the Jiaxi period of the Song Dynasty." The man who built this bridge, referred to here by his nickname Daogu, was none other than Qin Jiushao, the great mathematician of Southern Song.

Qin Jiushao (1208–1268) traced his ancestry to Fan County, now in Henan province, but historically located at the junction between Henan province and Shandong province. Since the county seat had been located in

Statue of Qin Jiushao (photograph by the author, Nanjing)

Shenxian County in Shandong province for hundreds of years, he referred to himself as a native of the state of Lu (now in Shandong). He was born and grew up however in Anyue County in the Puzhou, Sichuan, located between modern Chengdu and Chongqing. His father was a *jinshi*[1] scholar, who had at one point served as the local administrator of Bazhou (now called Bazhong, in northeastern Sichuan). According to the understanding

---

[1] Translator's Note: the *jinshi* was the highest degree of the imperial examination.

of this author, the people of his hometown claim that three generations of the Qin family were *jinshi* scholars. In 1219, a mutiny in Bazhou, in which his eldest brother was killed and his grandmother and sister-in-law died, prompted him to leave it behind and relocate to the capital Lin'An, in modern Hangzhou, where he and his family lived by Xixi Brook, a suburban area which had attracted various imperial officials and their families after the outbreak of a famous fire in Lin'An in 1201. This fire raged for 3 days and 3 nights, burning down the Imperial Ancestral Temple, the Three Departments and Six Ministries, the Censorate, and so on, affecting more than 36,000 families.

Qin Jiushao was bright and studious as a child and cultivated a wide range of interests. His father after coming to Lin'An served for a period as an official for the Ministry of Public Works and later as the Vice-Curator of the Imperial Library, which put him in charge of calendars for the subordinate Imperial Bureau of History; this gave him the opportunity to read widely and study astronomy, calendry, civil engineering, mathematics, poetry, and so forth. In 1225, he was appointed as prefect at Tongchuan (now Santai, in Sichuan), an important border area with Tubo, a Tibetan region of ancient China. He settled his family in Huzhou, not at all far from Lin'An, and took only his favorite and youngest son Jiushao with him to take up this post. Jiushao himself served later as a county military officer for Zhuoqi county (modern-day Qijiang in Santai county), for which reason he has also been described as a righteous leader of soldiers, capable of leading the troops in battle.

In 1232, Qin Jiushao likewise passed the *jinshi* examination and served subsequently as an official in Sichuan, Hubei, Anhui, Jiangsu, Hainan, Jiangxi, Guangdong, and still other places. He was called upon also to participate in military activities during the frequent outbreaks of war along the Jialing River basin (the Jialing River is a major tributary of the Yangtze River) after Yuan soldiers invaded Sichuan in 1236, a period of his life that he describes in the preface to his *Mathematical Treatise in Nine Sections*. He returned to Lin'An in 1238 to attend to the period of *ding you* for his father, an obligatory period of 3 years' mourning after the death of a parent in ancient China; afterward he moved to Huzhou for the same purpose. Noticing that there was no bridge across Xixi Brook, for which reason communication between the people on its opposite sides was quite inconvenient, he designed the bridge himself and obtained funding through friends from the treasury to see it built.

At the time, the bridge was not given any special name, and people typically referred to it as Xixi Bridge, after the river it spanned. It was only in the early Yuan Dynasty that Zhu Shijie (1249–1314), another outstanding mathematician and wide traveler from the north, visited Hangzhou and proposed

Daogu Bridge in Hangzhou (photograph by the author)

that Xixi Bridge be renamed Daogu Bridge in honor of its builder, whom he admired as a mathematician; Zhu Shijie personally engraved its new name upon the head of the bridge.

Daogu Bridge remained in this place, not far from the home of 19 years of the author of this book, until the turn of this millennium, although it is not clear whether or not it had ever been rebuilt at any point during the intervening centuries. Then, due to the expansion and reconstruction of Xixi Road, the original bridge and stream were razed and filled, respectively, and the Daogu Bridge Neighborhood Community that once existed disappeared as well. In their place were built high-rise buildings, including an International Business Center, the Zhejiang Provincial Department of Land and Resources, and Huanglong Century Apartments. There remained only a bus stop with the name Daogu Bridge, and allegedly a Daogu Bridge Road, not shown however on the maps.

In 2005, a stone pedestrian bridge was built across Yanshan Brook, a tributary of Xixi Brook at the south side of Tianmushan Road (the main east-west road in Hangzhou), more precisely between the Madeng Bridge on Hangda Road and the Yanshan Brook Bridge on Huanglong Road, about a hundred meters from the original site of Daogu Bridge and wider and more spacious than it. I visited this bridge on several occasions and saw that it crosses the river at a place of beautiful scenery, with weeping willows on either side, removed from the bustle of the city. Most importantly, the bridge had not yet been given any name, which inspired in me a sudden whim: I suggested for it the name Daogu Bridge. Later, I learned that Qin Jiushao was invited to design and rebuild Gao Bridge in Mingzhou (now Ningbo, an important port city) in 1256; at that time his friend Wu Qian was serving as the governor there.

Overview of Mathematics

In 1244, when Qin Jiushao was serving as *Tong Pan*, an official position in ancient China something like the vice governor of a state, responsible for such matters as the transportation of grain, household land, water conservancy, and litigation, in addition to monitoring the governance of the state, in Jiankang Prefecture (Nanjing), he had to leave his office and return to Huzhou in Zhejiang after the death of his mother in order to observe the 3-year mourning period. It was during this time of filial piety that Qin Jiushao turned his concentration in earnest to mathematics and completed in September of 1247 his masterpiece, the *Mathematical Treatise in Nine Sections*, a voluminous treatise of more than 200,000 words. He presented this work to Zhao Yun, Emperor Lizong of Song, to whose court he had been summoned on account of his great reputation, rich knowledge, and achievements in astronomy and calendry. He expounded there upon his opinions, introducing his work with the title *Overview of Mathematical Methods* or *Overview of Mathematics*. It can be said therefore that Qin Jiushao was the first Chinese mathematician to be received by the emperor. Some years later, the Hebei mathematician Li Ye was summoned three times by Kublai Khan.

The *Mathematical Treatise in Nine Sections* is divided into 9 chapters across 18 volumes, each chapter containing 9 questions, and surpasses comprehensively its predecessor, the classic *Nine Chapters on the Mathematical Art*. The most important of its contents are undoubtedly the *Da Yan Shu method* of the first chapter and the *method of positive and negative evolution* from the ninth chapter, concerning market transactions. The latter is also known as the *Qin Jiushao algorithm*, an algorithm for the numerical solution of polynomial equations of any degree, with positive or negative coefficients. The solution of such equations ordinarily requires an iterative method in which the polynomial is evaluated repeatedly, each such evaluation requiring $n(n + 1)/2$ multiplications and n additions for a polynomial of degree n. Qin Jiushao transformed this problem into a system of n linear equations, requiring only n multiplications and n additions to solve. He also worked out 21 examples, with polynomials of degree at most 10. This algorithm was only rediscovered in the nineteenth century by the British mathematician William George Horner, for which reason it is also known as *Horner's method*; its utility is by no means insignificant, even today in the age of computers.

The *Da Yan Shu* is a generalization and mathematically precise statement of Sun Zi's theorem, which dates back to a work entitled the *Sun Zi Suanjing*, written around the fourth- or fifth-century CE, in which there appears *the problem of the unknown number*: "Now there is an unknown number of things; if we count by threes, there is a remainder of two, if we count by fives a remainder of three, and if we count by sevens a remainder of two. What is the number?" The given answer is 23; in other words, Sun Zi presented only a special example of a general phenomenon. According to the folklore of

Huai'An, in Jiangsu province, this problem can be traced back to the famous story of the Western Han dynasty general Han Xin gathering his troops for a roll call sometime in the second- or third-century BCE:

> Apparently there was fighting in the final years of the Qin dynasty between the Chu and the Han, in the course of which Han Xin led his troops into battle against the Chu. The battle was a hard one, and the Han army suffered many casualties; in their retreat, they came to a hillside and suddenly received word that Chu army was behind them on horseback. There was dust in the distance and the murderous sounds shook the sky. The Han army was exhausted, but Han Xin ordered them successively to form into lines, first of three, and then of five, and then of seven. There were in the first formation leftover two soldiers out of line, in the second three, and finally two again, from which Han Xin concluded immediately "We have 1,073 warriors, and the ene:y is less than five hundred!" The morale of his army was greatly improved, and they turned to defeat the Chu army in a stroke.

We now describe the *Da Yan Shu method*, in the language of modern mathematics. Suppose $m_1, \ldots, m_k$ are pairwise relatively prime integers larger than 1, with product M. Then for any k integers $a_1, \ldots, a_k$ there is a unique positive integer x, not exceeding M, such that the remainders when dividing x by each of $m_1, \ldots, m_k$ in turn are given by $a_1, \ldots, a_k$, respectively. Qin Jiushao provides a detailed method for determining this x, making use of techniques known in modern number theory as the Euclidean algorithm (*Da Yan Shu*) and the solution of a certain linear congruence (*Qiu Yi Shu*), which requires the determination of x given relatively prime a, m with $m > 1$ such that the remainder of ax divided by m is exactly 1.

Unfortunately, because there was no concept of prime numbers in ancient China (and they were only recognized under this name during the Kangxi period of the Qing dynasty), and their use at the time was not theoretical, but turned up only in practical problems related to the calendar, engineering, taxation, and military affairs, Qin Jiushao did not write down a strict proof of the theorem he had discovered. On the other hand, this is less important with respect to theorems such as this one, which describe solution methods, and his work exhibited some subtle generality: he accounted for the possibility the relevant integers need not be pairwise relatively prime and gave a reliable algorithmic procedure for converting this case into the established case of relatively prime pairs.

Likeness of Song Lizong, the emperor of South Song Dynasty

In addition to this, Qin Jiushao discovered *Qin Jiushao's formula*, known to Western readers as *Heron's formula*, for determining the area of a triangle given the lengths of its sides. In the second chapter of the *Mathematical Treatise in Nine Sections*, entitled *Heavenly phenomena*, Qin Jiushao provided calendrical calculations and calculations of snowfall and rainfall. His statue appears among three statues of famous ancient meteorologists at the Beijige Meteorological Museum in Nanjing, with the inscription: "He used 'the number of rains on flat lands' (that is, the amount of rain per unit area) to measure rainwater, and was the first in the world to establish the scientific theoretical basis for the measurement of rainfall and snowfall."

European Awareness

In 1801, the *Da Yan Shu*, which is often referred to today as the *Chinese Remainder Theorem*, was rederived by Carl Friedrich Gauss, the prince of mathematicians in Part 2, Section 7, of his famous *Disquisitiones Arithmeticae*. The Swiss mathematician Leonhard Euler had also already carried out extensive research in this direction. Neither of them knew that this result had already been discovered by Chinese mathematicians. In fact, it was not until 1852 that the results and methods of Qin Jiushao were translated and introduced to Europe by the British missionary Alexander Wylie (who also collaborated with the Qing dynasty mathematician Li Shanlan on the first complete Chinese translation of Euclid's *Elements*) in a paper entitled *Jottings on the Science of the Chinese*, which received widespread attention in European academic circles and was quickly translated from English into German and

French. The Qin Jiushao algorithm appeared in this article as well. As for when and how it came to be known as the *Chinese Remainder Theorem*, this remains a mystery, but it seems to have happened no later than 1929.

Strictly speaking then, this theorem, also known as Sun Zi's theorem, should be called Qin Jiushao's theorem, as I did in my book *A Modern Introduction to Classical Number Theory* (in both its Chinese and English editions). The story of the name *Sun Zi's theorem* is related to a certain moral stain attached to Qin Jiushao, as we discuss below. According to the analysis of Pan Chengdong, a predecessor of this author, there is a certain dismissiveness in the use of the name *Chinese Remainder Theorem*, indicating the focus of ancient Chinese mathematicians on calculations and their lack of theoretical achievement. In any case, this is certainly the most influential theorem discovered by the Chinese, and it cannot be avoided in any basic textbook in number theory. Moreover, it has a natural and important generalization to abstract algebra and important branch of modern mathematics. It also has wide applications in cryptography, the calculation of polynomial interpolations for numerical analysis, the proof of Gödel's incompleteness theorem, the negative solution of Hilbert's tenth problem on the solvability of Diophantine algorithms, the fast Fourier transform, and in many other fields.

The German historian of mathematics Moritz Cantor praised Qin Jiushao as the luckiest genius for discovering this theorem, which had not yet been given a name in the West at the time; there are echoes in this of the praise that the great French mathematician Joseph-Louis Lagrange paid to Newton, observing that there is only one opportunity to discover the law of gravity. The Belgian-born American historian of science George Sarton wrote of Qin Jiushao that he was "…one of the greatest mathematicians of his race, of his time, and indeed of all times." Of the 12 mathematicians highlighted in the book *History of Mathematics from Mesopotamia to Modernity* (published in 2005 by the *Oxford University Press*), Qin Jiushao was the lone Chinese mathematician.

The anecdote of Qin Jiushao and his bridge building recall another story concerning Isaac Newton. There is a bridge today over the River Cam, which flows through Queen's College at Cambridge known as the Mathematical Bridge. This is because the designer of this bridge is said to have been none other than Newton himself, working in the seventeenth century and apparently constructing the bridge without the use of a single nail. At one point, the bridge was quietly taken down in order to test this claim, but afterward, it could not be put back together again, and instead a new bridge was built in its place. In any case, the Mathematical Bridge has long been a popular scenic spot and essential destination for tourists visiting the university.

The Mathematical Bridge in Cambridge, said to have been designed by Newton (photograph by the author)

The story of Daogu Bridge is even older than this one (some four centuries earlier than Newton) and involves two great ancient mathematicians, although this was little publicized and even among mathematicians working in Hangzhou it was not well known that the Daogu Bridge of ancient times refers to Qin Jiushao. In 2012, the Hangzhou Municipal Government accepted my suggestion to rename the stone pedestrian bridge not far from the original Daogu Bridge, so that Hangzhou can again have a new Daogu Bridge to contribute to the scientific and cultural landscape of this historical city. I asked the mathematician, Mr. Wang Yuan, to inscribe the name of the bridge in a stone tablet set up at its head.

Moral Stains

It must be pointed out that in his later years, and to subsequent generations, Qin Jiushao became a controversial figure. His academic achievements were not recognized by his contemporaries, and rumors swirled around him of corruption, perversions of the law, extravagance, and even crimes against human life to the extent that he could no longer be considered human. There are no biographies of Qin Jiushao in any of the Song histories or local annals, his name and the name of the bridge appear and disappear, and the whereabouts of his descendants are unknown. His name does not appear in Chinese number theory textbooks, and only the likeness of the ancient mathematician Zu Chongzhi adorns Chinese campuses in portraits and sculptures. Although Qin Jiushao is the only Chinese mathematician to appear in the four-part documentary *The Story of Maths* produced and aired by the BBC, and the most extensively treated among seven mathematicians of the East to appear in that series, there too the heights of his academic achievement are set in contrast against exaggerations of his moral stains.

After many inquiries and consultations, the author learned that there were two main sources for rumors concerning Qin Jiushao, with very similar contents. The Fujian poet Liu Kezhuang, who was 15 years older than Qin Jiushao, wrote the *Report on the Appointment of Qin Jiushao as Military Governor of Linjiang*, which appears in *The Collected Works of Mister Houcun*; and in *Miscellaneous News from Guixin* there appears an article *Qin Jiushao* written by Zhou Mi, a Huzhou scholar some 30 years younger than Qin Jiushao (this article is two pages in total, and the title of the collection refers to Guixin Street in Hangzhou, where Zhou Mi stayed after the collapse of the Song dynasty). The latter was listed under the category *novelists and the like* in the *Siku Quanshu* of the Qing dynasty Qianlong period. Both Liu Kezhuang and Zhou Mi could be described as people of some influence and status in the literary history of the Southern Song dynasty.

During the Jiaqing period, Zhou Mi came under criticism for slander and the spread of rumors from Jiao Xun, a famous scholar of the Yangzhou school in both the liberal arts and sciences, from Ruan Yuan, a Confucian scholar and educator who served as governor of Zhejiang, and later from Lu Xinyuan, a Huzhou scholar and bibliophile of the late Qing dynasty, and the reputation of Qin Jiushao improved to the extent that someone began to work on a biography. Similarly, Liu Kezhuang, a follower of Jia Sidao, was subject to ridicule in his own time for flattery and the excessive promulgation of his numerous writings. In 1842, the *Mathematical Treatise in Nine Sections* was revised for the first time by Song Jingchang, a famous master of calendars, and published in print for the first time, ending a history of some six centuries of copying. Its manuscripts had previously appeared in the *Yongle Dadian* of the Ming dynasty and *Siku Quanshu* of the Qing dynasty. During the process of copying and dissemination, it gained at one point the title *Nine

Chapters on Mathematics and later was given its current title by Wang Yinglin, a dramatist of the late Ming dynasty. This history of transmission invites comparisons between the *Mathematical Treatise in Nine Sections* and the epics of Homer.

Jia Sidao, with whom Liu Kezhuang associated himself, was a treacherous chancellor during the reign of Emperor Lizong of Song; by contrast, Qin Jiushao was associated with the chancellor Wu Qian, who had placed first in his imperial examination and was famous for his integrity and selflessness, his loyalty and patriotism, and his concern for the country and its people. He was also a poet, an expert in water conservation, and a hero of the conflicts with Japan. Unfortunately, Jia Sidao was the younger half brother of Emperor Lizong's favorite concubine; he became known as the cricket chancellor. In 1261, Jia Sidao made accusations against Wu Qian, who was nearly 70 years old at that time, and the latter was dismissed and exiled to Xunzhou (now Longchuan in Guangdong), not far from Meizhou (Meixian), where Qin Jiushao served as military governor. He succumbed to a sudden illness and passed away the following year (it is suspected that he was poisoned), while Qin Jiushao went to attend his funeral and continued to serve in Meizhou until his death there in 1268, according to Zhou Mi.

Wu Qian was native to Ningguo in Anhui, adjacent to Huzhou, although some accounts claim him as a native of Deqing in Huzhou instead. He had met Qin Jiushao during his time in Bazhou, and they later lived together in Huzhou during their periods of *ding you*. It is possible that the house Qin Jiushao built atop a foundation contributed by Wu Qian was somewhat extravagant, inviting the jealousy of the local literati. Notably, Zhou Mi was local to Huzhou, in contrast to the foreigner Qin Jiushao. In any case, Zhou Mi was talented as a poet and writer and also sustained interests in history, painting, medicine, and even arithmetic. His *Zhiyatang Miscellany* is a collection of jottings on various topics. Perhaps most intriguingly, it contains an account of the troop arrangements of Han Xin, referred to as *Ghost Valley Calculation* or *Wall Partition Calculation*.

On the other hand, Qin Jiushao was also an accomplished poet and had received guidance from the famous poet Li Liu, so the possibility of relative indifference between the literati cannot be ruled out. His experience was almost certainly the product of political strife. He spent the last 6 years of his life in Meizhou; in fact, when the Jin people and the Mongolian army made their excursions to the south, Wu Qian and Jia Sidao represented, respectively, the war faction and the peace faction. The author once made a special trip to Meizhou. It is surrounded by mountains, and the streets of the city are filled with banyan trees. I thought of Cai Lun, the inventor of papermaking. Undoubtedly, his achievement alongside those of Qin Jiushao represents the apex of science and technology in ancient China. Cai Lun earned a title for his work in papermaking, but after the death of the Empress Dowager, who had doted upon him, he was poisoned by her political enemies.

There are some interesting remarks written later by an unknown author in *Zhang Xiang Si*: "Last autumn, this autumn, the people by the lake were happy and then sad, West Lake still flows, Wu Xunzhou, Jia Xunzhou, passed fifteen years and let go of life," meaning that 15 years afterward, Jia Sidao was demoted in station to the same point as Wu Qian, who later died there, killed en route by a prison escort with a sense of justice. During the Qing dynasty, Wu Lu, a descendant of Wu Qian, honored his ancestor and took again the first place in the examination. Most of the Wu family lived in Fujian, and the Wu Ancestral Home on Tumen Street in Quanzhou is also known as Dongguan Xitai. As for the final years, burial place, and descendants of Qin Jiushao, although the author has endeavored to find something out in Meizhou, there remains no way of knowing. If it were really the case as Liu Kezhuang and Zhou Mi had written in their articles that Qin Jiushao had long been worthy of execution or imprisonment, how could it have happened that he received instead only a demotion and continued to govern?

There is a famous illustration worth mentioning in the *Mathematical Treatise in Nine Sections*, which is used to calculate the height of a pagoda spire. The calculation is carried out by recording the angle of observation and applying the tangent function. The shape of the pagoda is similar to that of the Feiying Pagoda at Feiying Temple in Huzhou City, dating to the Tang dynasty. This double pagoda consists of an inner stone pagoda and, outside a pagoda of brick and wood, a unique structure of a pagoda within a pagoda, designated as a Major Historical and Cultural Site Protected at the National Level. The inner and outer towers were built in the Tang dynasty and Northern Song dynasty, respectively, but they were demolished around the twelfth century, and the current tower was rebuilt in the 1330s, just before Qin Jiushao settled in Huzhou.

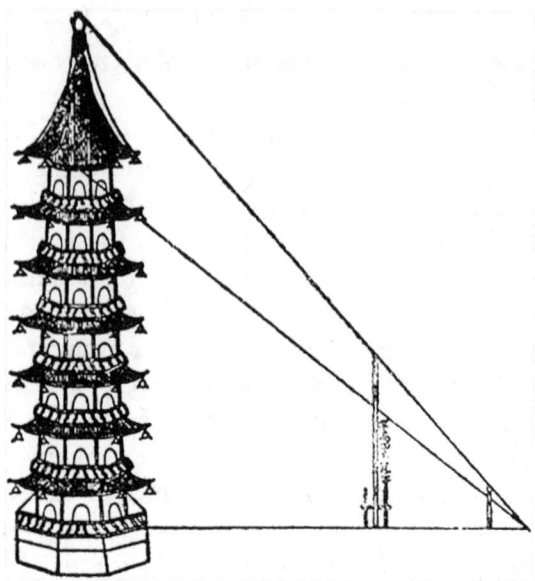

Illustration from the *Mathematical Treatise in Nine Sections*, said to depict Feiying Tower in Huzhou

The Origins of the World

Finally, I would like to talk about the preface to the *Mathematical Treatise in Nine Sections* and the prefatory poems placed at the start of each chapter. The first volume deals with *Da Yan Shu*, or the invaluable theory of linear congruences. In the preface to this chapter, Qin Jiushao wrote: "The lofty Kunlun mountains, majestic and powerful; the origins of the world lie in mathematics." The second chapter concerns the astronomical calendar and opens with the words: "The seven planets are swirling in the sky, and everything in the world is constantly changing." The third volume, about the measurement of land, with the words: "The common people are small, but they should be put first; assess the situation to see the world."

The remaining chapters cover surveying, transportation taxation, storage of grains, construction of buildings, military matters, and pricing and interest in transactions. In the chapter on taxation, Qin Jiusho wrote: "Officials must

4 Qin Jiushao, Daogu Bridge, and the *Mathematical Treatise*...

Meizhou, Guangdong, where Qin Jiushao died (photograph by the author)

implement benevolent policies, consider the people, and put themselves in their shoes, as if they were starving and suffering through disasters. If taxes and labor are unevenly distributed, how can the people feel at ease?" From this it can be seen that in his works, Qin Jiushao considered not only mathematics but also natural phenomena and social life, providing an important reference for future generations to understand the social, political, and economic life of the period.

The prefatory poem to the chapter concerning grains even points out some of the social ills of the times, with well-reasoned foundations: "The taxation of grains depends on the grade, and the storage of grains depends on the season; every grain of millet, every inch of silk is the result of the labor of men and women," "High-ranking officials compete with one another and bully the people; corrupt officials both large and small scheme with all their might," "The management of finances should be as a wise man manages water: cleaning the source, order instead of chaos, treating symptoms and root causes,

Qin Jiushao's Museum (photograph by the author, Anyue, Sichuan)

eliminating hidden dangers," "Stupid officials turn a blind eye, the people are miserable, the torture goes on. Further and further from reason, the unkind official is so deplorable!"

At the start of the general preface, Qin Jiushao mentions that mathematics was one of the six arts (*liu yi*) of the Zhou dynasty (the others being etiquette, music, equestrianism, archery, and calligraphy). Scholars and officials from ancient times onward had always respected and advocated for mathematics, which came into being because people want to grasp the laws of the world. From the largest perspective, mathematics offers understanding of life and nature; from a smaller perspective, it assists with the management of affairs and the classification of all things. Qin Jiushao believed deeply that everything in the world is related to mathematics, a system of thought he shared with the Pythagoreans of ancient Greece.

It was this that attracted Qin Jiushao to mathematics. He sought the guidance of scholars and capable people and explored profoundly its subtleties: "In my youth I was living in the Lin'An with my father, so that I was able to study at the National Astronomical Observatory; subsequently, I was instructed in mathematics by a recluse scholar [*yin junzi*, which here does not indicate an addict but rather someone who has escaped the world to live in seclusion]." At that time, the Yuan army had invaded Sichuan, and Qin

Jiushao was obliged to travel long distances during the war but never abandoned his mathematical studies.

Simultaneously, Qin Jiushao lamented the status and low regard of mathematicians among the people of his day. By this he meant mainly pure mathematics and methods and techniques that temporarily have no apparent application. He felt that mathematics was an object of contempt mathematicians viewed only as tools, similar to a maker of musical instruments who only produces their sounds: "Originally I wanted to promote mathematics to the level of philosophy (*tao*), but it was simply too difficult." All of this suggests a thoughtful person of taste, difficult to reconcile with the Qin Jiushao of infamous reputation. After more than seven centuries, the moral stains surrounding Qin Jiushao remain the biggest mystery in the history of ancient Chinese science.

Hangzhou, May 2012

5

The Reclusive Frenchmen: Descartes and Pascal

Our most excellent people study mathematics.
—*Paris citizen*

From the Provinces to Paris

To borrow a phrase from Voltaire, the seventeenth century in European history was the century of Louis XIV, and it can be regarded as the century in which France emerged as a great power. Turning to the history of science, Alfred North Whitehead called the seventeenth century the century of genius. This century saw France contribute three mathematical geniuses: René Descartes, Pierre de Fermat, and Blaise Pascal. It is well known that Fermat was mainly interested in pure mathematics, and he is famous in particular for Fermat's last theorem, which remained unsolved for many years until it was finally conquered at the end of the last century. Descartes and Pascal on the other hand were versatile thinkers, both of whom lived in Paris for a time, and so they became famous already during their lifetimes.

It is interesting that all three of these great minds were born in further flung provinces (or departments as they are known in French) around the same time: Descartes was born in Indre-et-Loire in the midwest, Fermat in Tarn-et-Garonne in the south, and Pascal in Puy-de-Dôme in central France. These departments, among 96 in total in France, are 300–1000 km from Paris and host no famous cities; the three figures under discussion were born, respectively, in a village, a small town, and a provincial capital. This seems to confirm my earlier conclusion that metropolitan cities do not easily produce geniuses.

To maintain the focus of our topic, we leave Fermat aside and focus only on Descartes and Pascal, two geniuses spanning the fields of science and the humanities, with similar childhood experiences. Both lost their mothers at an early age, and both were frail and sickly children. Fourteen months after Descartes was born, his mother died of tuberculosis, which disease she passed on to him. Pascal, who was also frail, lost his mother when he was 3 years old. On other hand, both had wealthy and respected fathers, although with totally different attitudes toward their children.

Like Fermat, Descartes' father was a consultant to the local parliament. After the death of his wife, he moved to another country and remarried, leaving behind his son to be brought up by his maternal grandmother. They rarely saw one another after that, but his father was financially generous, which enabled Descartes to receive a good education and afforded him the opportunity to enter the aristocratic school established by the king. Upon graduating, Descartes studied law at the University of Poitiers, further from Paris. Three years later, uncertain about his career path and wanting to see the world, Descartes joined the Dutch army and later transferred to Germany.

Likeness of Descartes

5 The Reclusive Frenchmen: Descartes and Pascal

Likeness of Pascal

When he was 26 years old, Descartes sold the property left to him by his father. This allowed him to live comfortably, and from that point on he was able to freely do as he chose. He first spent 4 years traveling in Europe, spending 2 years in Italy, before settling in Paris. This was the same year, when Descartes was 30 years old, that Pascal's mother died, leaving behind three young children. Fortunately, his father, a Latin scholar and mathematician who was responsible for the discovery of Pascal's spiral, had a kind heart. He retired early from his job presiding as a judge over tax cases at court and moved his family to Paris in order to see to the education of his children, in particular his frail and very delicate son. He never remarried.

The teaching methods of his father emphasized problem-solving rather than rote communication of materials, prompting young Pascal to develop a curious and adventurous spirit, with an outstanding talent for hands-on experimental work. In light of his son's physical condition, however, his father focused mainly on language education, teaching him only some basic principles in mathematics, which actually led his son to become more curious about and sensitive to the topic. Pascal is said to have independently derived the theorem that the sum of the three interior angles of a triangle is equal to two right angles at the age of 12.

From that point on, his father began to teach him Euclidean geometry, and soon they were both participating in the weekly mathematical salons hosted by the priest Marin Mersenne, which was the prototype for the French Academy of Sciences. Mersenne was an indispensable member of the French mathematical community of the seventeenth century. He organized salons and secret trips, bringing him into close and fruitful contact with his three most distinguished and distinctive colleagues, Descartes, Pascal, and Fermat. He also left his own mark on the history of mathematics via the Mersenne primes.

By contrast, Descartes developed an interest in mathematics relatively late, while he was a soldier in the Netherlands, where he saw the solutions to mathematical questions written in Flemish on the bulletin board of the military camp. Another soldier, who at that time was translating for him, was accomplished in mathematics and physics and soon became his mentor. Four years later, Descartes wrote to this comrade that it was he who had awakened him from his indifference and excitedly communicated that after 6 days of intense mathematical work he had managed to make four important discoveries.

Achievements in Math and Science

Descartes was clever and realized early on that the essence of mathematical methods is to begin from propositions, those the truth of which can be clearly discerned through intuition and from which other propositions can gradually be derived deductively. That is to say, he took into account the internal rigor of mathematics, which can be extended to the sciences, without neglecting the perception of the senses. In this way the shackles of longstanding authority can be discarded, liberating people to engage in self-inquiry, a nonempirical approach that serves as an alternative to the formal syllogisms of Aristotle, and instigating a new zeitgeist.

5 The Reclusive Frenchmen: Descartes and Pascal

As Descartes pointed out, syllogistic laws are useful only for communicating what we already know, not for discovering what we do not. But because of the authority of scholastic philosophy, the methodology of Descartes spread slowly through Western Europe through informal channels. It reached Newton in England while he was still studying at Cambridge University. Later he was inspired by a fallen apple at his family farm to conceive and formulate the law of universal gravitation, a result of taking to heart the new ideas of Descartes. In mathematics, Descartes achieved his greatest results in geometry.

Today we have no way of knowing which exactly were the four major discoveries that Descartes mentioned in his letter to his mentor. In the appendix to his *Discourse on the Method*, which was deliberately delayed due to the conviction of Galileo by the Inquisition, Descartes presented some geometric discoveries, including the classification of conics, the method of tangents to curves, and the solution of higher-order equations, all of which were outdated. From the point of view of the author, the main mathematical contributions of Descartes to mathematics are summarized by the following four points:

- First the symbolization of arithmetic, including the introduction of the letters $a, b, c, \ldots$ for known quantities and $x, y, z, \ldots$ for unknown quantities and exponents
- Second, the extension of the x-axis and y-axis from a given origin, establishing for the first time in history an oblique coordinate system, providing an example of a rectangular coordinate system, and establishing analytic geometry
- Third, discovering the relationship between the number of vertices v, edges e, and faces f of a convex polyhedron, which can be stated symbolically as $v - e + f = 2$, later known as the Euler-Descartes formula
- Finally, the introduction of the curve known as the folium of Descartes, which appears frequently in calculus tutorials to this day

It is readily apparent that Descartes derived his passion for mathematics from his need for methodology. He felt that knowledge requires certainty, which is precisely what mathematics provides. So after a brief period of enthusiasm, Descartes turned his attention to the broader problem of looking for solutions to problems across all of science. In fact, he at one point had high hopes for mathematics. Much as Pythagoras had loved natural numbers and famously believed that everything is number, Descartes believed that any problem could be made mathematical and any mathematical problem could be solved algebraically, reducing all problems to a matter of equations.

In contrast with the theoretical importance that Descartes attached to intuition and deduction, Pascal approached mathematics more through experience and practice. At the age of 17, he published his first article, the *Essay on Conics*, which was soon lost and did not resurface until more than a century had passed. This article proved several esoteric results in projective geometry, including what is known today as Pascal's theorem. This theorem states that the three intersection points of the three opposite sides of any hexagon inscribed in a conic section lie on a single straight line. Descartes ridiculed this work, but it has proven to be one of the most fruitful results in all of geometry.

Two years later, in an attempt to simplify the calculation of taxes for his father, who had returned to Rouen to serve as its governor, Pascal undertook and successfully completed the development of a mechanical calculator, the first computer created by humankind. Although cumbersome, this calculator was capable of arithmetic calculations with eight-digit numbers. Over the following 10 years, Pascal continued to improve and perfect his device, eventually building more than 50 calculators, which, however, he was not very effective at advertising and bringing to market; today, one of the eight of these machines that still exist is owned by IBM, and in the early 1970s a computer language developed in the United States was named after Pascal in memory of his achievements in mechanical computation.

Shortly after his passion for calculators came to an end, one of his noble friends who adopted the title of *knight* put to Pascal some questions concerning the odds of winning or losing in gambling, to which he was addicted. This led to a deep investigation, which Pascal carried in frequent communication with Fermat, far in the remote southern mountains. Historians of mathematicians generally attribute to this correspondence the foundations of probability theory, today an independent branch of mathematics. This also prompted him to formulate an argument known today as Pascal's wager, which he discussed in his most important prose work, the *Pensées* (or *Thoughts*), in one of its longest and most famous fragments. The premise of this argument is that the propositions that either God exists or God does not exist have the same complementary form as a wager.

As a by-product of his studies in probability, Pascal also determined the relationships between binomial coefficients. These coefficients, arranged in a triangle according to ascending powers, are now referred to as Pascal's triangle in the West, and the relationship between them constitutes one of the basic results in combinatorics. In fact, this triangle diagram appears already in the works of the Northern Song dynasty mathematician Yang Hui, who attributed them to a lost work of Jia Xian, who lived more than two centuries earlier. Therefore, Chinese textbooks refer to it either as Jia Xian's triangle or Yang Hui's triangle; it is not known today whether Jia Xian discovered and demonstrated it on his own.

5 The Reclusive Frenchmen: Descartes and Pascal

Both Pascal and Descartes made outstanding contributions to other sciences in addition to their mathematical contributions. In fluid mechanics, there is Pascal's law, which states that if the pressure at any point of a confined incompressible fluid changes, it will be transmitted equally throughout the fluid and to the enclosing walls; moreover, the pressure is given by the force divided by the area of action. The pascal is also an international unit of pressure; it occurs, for example, in weather forecasts, describing say the number of kilopascals at the center of a tornado or hurricane (the lower the pressure, the stronger the wind). In addition, the first mathematical formula to play a part in the insurance industry is also attributed to Pascal.

Descartes had broader interests, in which respect he was later followed by Goethe, and involved himself in optics, meteorology, and physiology. In meteorology, for example, Descartes endeavored to explain the phenomenon of rainbows via his refraction theory of light and presented an analysis of color

The folium of Descartes

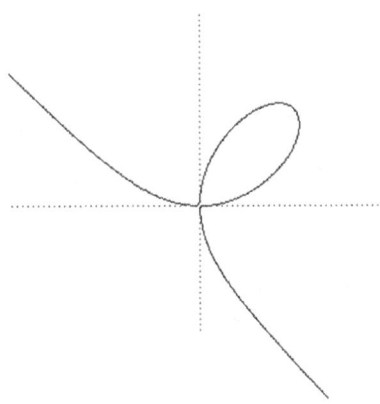

Folium of Descartes

Pascal's triangle

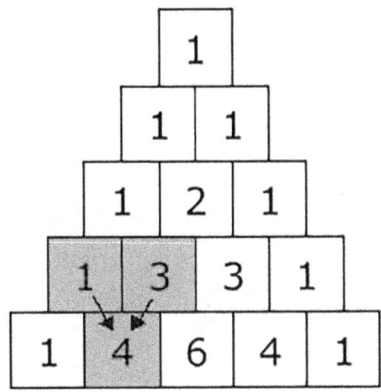

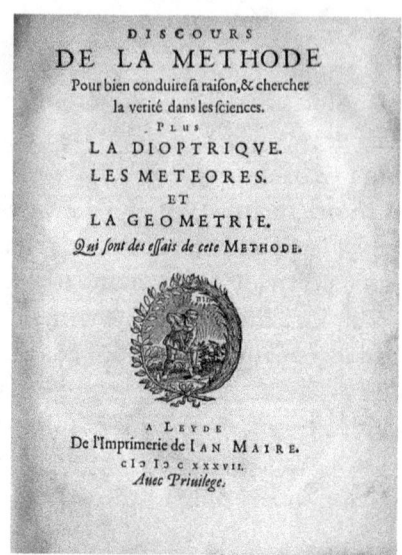

Title page of *Discourse on Method* by Descartes; *The Geometry* is the third appendix

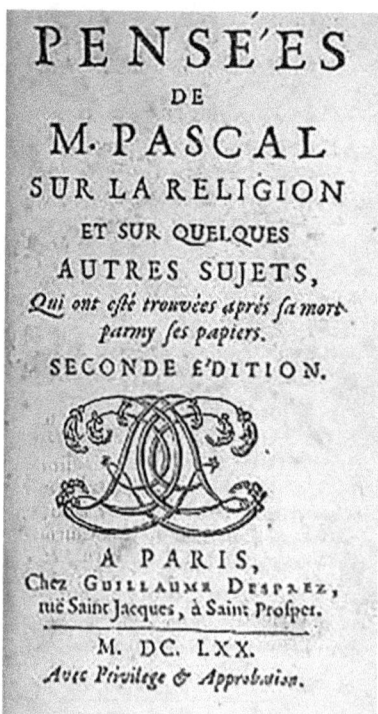

Title page of the second edition of Pascal's *Pensées*

5 The Reclusive Frenchmen: Descartes and Pascal

according to the rotation speed of elementary particles. In physiology, his ambitions rose even higher, and his considerations included digestion, respiration, blood circulation, and the nervous system. He also studied the principles underlying sneezing, coughing, and yawning, the mechanisms of perception, and the lens of the eyeball and even purchased corpses for the purposes of dissection. His description of the heart however came under criticism by William Harvey, the British doctor who discovered the correct theory of blood circulation. Bertrand Russell has pointed out that Descartes did far less outstanding work in these subjects than he did in mathematics and philosophy.

Peculiar Skepticism

In all of Descartes' writings, we can observe his efforts to find the uniting feature of knowledge, his belief that the various sciences are at once independent and interconnected. His book *The World*, for example, includes natural theories related to such topics as heat, light, the tides, the formation of the earth, and the characteristics of comets. He tried to synthesize all knowledge into a certain system or science based on a few simple principles, with the purpose of making nature more clear. He wrote in his notes that if we can understand how the sciences are connected, it will be easy to see that mastering them is no more difficult than memorizing a sequence of numbers.

Pascal did not entertain such ambitions, perhaps because he remained under the care of his father since childhood and maintained very close relationships with his two sisters, who cared for him meticulously in both material and spiritual aspects. His health was also poorer than that of Descartes. When he was young, doctors advised him not to engage in too much mental work, and for medicine he turned to recreation and entertainment, pursuing a course of treatment in dancing, sport, hunting, and gambling, which brought him into contact with fellow gamblers and carry out research into games of chance. While his talent can be said to have made him invincible, it also held him back from considering the unification of mathematics and natural science.

As they aged, Descartes and Pascal both found their interest turning from the material world to the spiritual. Descartes completed his *Discourse on the Method*, *The World*, *Meditations on First Philosophy*, and *Principles of Philosophy*, while Pascal left behind his *Lettres provinciales* and *Pensées*. One difference is that Descartes indulged more in metaphysical abstractions, due to the trial and conviction of Galileo, which served philosophy well but not science, whereas Pascal, who was deeply religious and starved for love, wrote with deeper piety and affection between the lines.

We have mentioned already at the beginning of the previous section that it was Descartes who first liberated philosophy from the shackles of traditional scholasticism. Hegel and his other philosophical successors revered him as the father of modern philosophy. A thoroughgoing dualist, Descartes drew a sharp distinction between mind and body, with the role of the former encapsulated in his most famous philosophical proposition: *I think, therefore I am.* This extraordinary expression has been later called into question by Russell and others but remains one of the most powerful formulations in the history of philosophy. As for chronology, the first step was the use of two-dimensional coordinate geometry to create analytic geometry, on the basis of which Descartes created his dualistic philosophical system.

Pascal had a fuller understanding of human limitations and became aware early on of human fragility and faults. Just as Descartes was searching for certainty through mathematics, Pascal intended through his thinking about the world to overcome inner anxiety and attain a kind of certainty. This pursuit shook him to awe and shock when he turned his consideration to ideas of the infinitely small and infinitely large. Whereas his mathematical discoveries lived in finite spaces, in talking about the universe he wrote that "the eternal silence of these infinite spaces frightens me," whereas in comparison even with the smallest mite, the infinitely small is like a new abyss.

Pascal was devoutly religious and originally envisioned the title *Apology for the Christian Religion* for his *Pensées*, but this work was published only after his death and its familiar name was chosen by its editor, whose decision has since proved fortuitous, since no doubt its influence would have been greatly reduced. This is one of the masterpieces of French literature and includes the famous wager discussed already: if God does not exist, there is nothing to be lost in believing in him nevertheless, but if he does exist then there is eternal life to be gained from belief.

In a letter written after the death of his father (collected in *Thoughts, Letters, and Minor Works*), Pascal writes that, without Christ, death is "horrible, detestable, and the horror of nature. In Jesus Christ, it is altogether different; it is benignant, holy, and the joy of the faithful." But nevertheless he viewed skepticism as a necessary precursor to faith. Naturally, his skepticism took a more constructive than destructive turn; its basic criterion was that any true religion must explain the human condition better than its competitors.

Descartes, too, was religious and went to great lengths to rigorously prove the existence of God along various lines of argument; in mathematics we can draw an analogy to the obsession of Gauss with the law of quadratic reciprocity in modular arithmetic. Descartes did believe himself to have completed the proof, but with less elegance than he had achieved in mathematics. Although the prose style of his argument was charming and personal, Russell

5 The Reclusive Frenchmen: Descartes and Pascal

University of Clermont-Ferrand II, also known as Blaise Pascal University

University of Paris V, also known as René Descartes University (photograph by the author)

has pointed out that it is essentially a scholastic argument (Russell himself was not at all religious). Moreover, one of the ontological arguments later came under severe criticism by Immanuel Kant.

Descartes believed that the human mind is basically sound and the only means through which to obtain truth. His attitude toward God is therefore suspect, and it is possible that like Michel de Montaigne he kept his faith only for reasons of custom. In his general thought and methodology, Descartes was otherwise purely a skeptic, and he viewed doubt as a necessary means and tool in philosophical and psychological inquiries, observing that we sustain many prejudices from childhood that carry through into adulthood if not corrected. He wrote that doubt is the art that liberates us from the influence of feeling.

Anonymous Gentlemen

Although they were separated by 27 years of age and it was only for a very limited time that they were both in Paris, nevertheless Descartes and Pascal stood to one another in a kind of rational and intellectual rivalry. On an autumn day in 1647, Descartes, who was already famous, paid a visit to young Pascal and expressed his admiration for the mechanical calculator the latter had manufactured. He also passed along some medical teachings and advice to Pascal, who was suffering from illness at the time. It is generally believed that this was the only time the two of them ever met.

On the other hand, Descartes disagreed with the experiments and research that Pascal had carried out on the existence of the vacuum, which he did not consider to be real; Descartes emphasized that it is not experiment that must play the decisive role in the discovery of truth. The facts have since vindicated Pascal, showing that the vacuum does indeed exist under certain conditions. Pascal criticized in turn some of the philosophical concepts put forward by Descartes, in particular his overreliance on science, arguing that reason is not enough to understand life: "The heart has its reasons which reason knows not." This should have come as no surprise to Descartes, who observed in his travels that customs vary from place to place, just as do the opinions of philosophers.

In any case, the dispute between Descartes and Pascal took place entirely within the realm of rational discourse and scholarship. In fact, neither of them was drawn to publicity. Descartes spent most of his life in the Netherlands, a country much of which lies below sea level, and was unwilling to disclose freely even to some of his close friends exactly where he lived; in fact, he changed his residence again and again, firm in the belief that deeper seclusion leads a better

5 The Reclusive Frenchmen: Descartes and Pascal

life. Pascal never attended school, nor sought any public office. Probably the place he visited most often in his life was the salon of Marin Mersenne, and even his frequent social outings were due to the advice of his doctor.

Here it is worth mentioning that in France in the seventeenth century, salons such as that of Mersenne where science and philosophy were freely discussed were very popular among the upper class. Whereas the cultural media in modern China (following in the footsteps of the United States) often focus on the stars of business and entertainment, the center of attention for Parisians of that era was rational life. It was no doubt due to their experience of infinite scenery and this enviable life that Descartes and Pascal turned to lives of seclusion. Somewhat earlier than both of them, Montaigne also sold off his official position at the age of 37, withdrew from social life, and returned to his manor.

Just after the turning of the New Year, 1655, when Pascal was 31 years old, he followed his younger sister to the Jansenist convent at Port-Royal in the southwestern suburbs of Paris. Subsequently, he composed papers only at the request of others and never published anything, including mathematical papers, under his own name. His two prose masterpieces were written while in seclusion there. One night, suffering unbearably from a toothache, he studied the laws of movement of the cycloid and produced a series of results. The cycloid is the curve formed by the trajectory of a fixed point on the circumference of a wheel rolling in a straight line along a flat road; it has been called the Helen of geometry.

In a sense, the lives of Descartes and Pascal represent two extremes; the one lived with total independence since his childhood, while the other was doted upon by his family members; one traveled around Europe, and the other never ventured outside of France. The two women in the life of Descartes, and the daughter who died, were open secrets, while we can only speculate on the basis of clues in his prose works as to whether or not Pascal ever enjoyed romantic love. But any genius needs only a brief period of passionate love, in the best case an unattainable one.

There are still two essential moments in the lives of these two figures that must be mentioned: the visions of Descartes and the conversion of Pascal. One night, early in the winter of 1619, when Descartes was stationed with the army in Ulm, Germany (the birthplace of Einstein, which also gives its name to the street on which the École normale supérieure can be found in Paris), he experienced a series of visions or dreams which provoked a revelation with respect to his mission in life: to unify all of knowledge according to the principles of geometry. It was this that prompted Descartes to sell off the property he had inherited from his father and concentrate entirely on his intellectual passions.

Pascal underwent two conversions, separated by 8 years. The first convinced him to join with his family in the fatalistic theological movement of Jansenism, while the second caused him to abandon his earlier designs and take his place at Port-Royal. These two conversions were both caused by coincidental accidents. As a result, although he never completely abandoned scientific research, it was no longer at the center of his heart; on the other hand, French literature gained two masterpieces. Pascal was recognized as a skilled rhetorician since childhood, with a talent for humor, and this no doubt played a part in the transmission of his work to future generations. The same is true of Descartes, who gained recognition as the inventor of modern philosophy, which cannot be separated from those dreams.

When Descartes reached his middle age, his 5-year-old daughter died of fever; his mistress, who had been living with him, married, and his period of happiness came to an abrupt end. He may have later fallen in love with an aristocratic lady more than 20 years his junior and suffered through a time of inescapable mental torture, until another noblewoman appeared in his life: Christina, Queen of Sweden, who sent a warship to invite him to Stockholm, where, despite a particularly cold winter, the French thinker who had loved sleeping in since childhood could come three times a week to the palace early in the morning to teach her philosophy. Several months later, Descartes died far from his home country due to a recurrence of pneumonia.

René Descartes commemorative stamp

Blaise Pascal commemorative stamp

"Man is only a reed, the weakest in nature, but he is a thinking reed." Pascal wrote these words at Port-Royal, where he maintained an extremely strict asceticism. When he found himself too given to talking, he had a belt covered with nails tied around his body as punishment. But his was a philanthropic heart, and even in the final year of his short life, he personally designed the first carriage line for the citizens of Paris and later on suggested that the government establish a company for the operation of this new mode of transportation, the predecessor of the bus and taxi service companies in every city in the world today.

Prior to Descartes, the French nation had made no outstanding achievements in the fields of science and philosophy. We may say that it was Descartes who opened up the paths of reason and the intellect to the French, just as Leibniz later did for Germany. As for Pascal, the French economist Jacques Attali observed in his biography of Pascal that he not only contributed to the flourishing of the French language but also to the full development of the French ethos, which became through his writings obscure and complex, a mixture of rationalism, antagonism, and universality. Briefly stated, France became France due to Descartes and Pascal, these two contemporaries and intellectual rivals.

January 2006, Hangzhou
Revised February 2015

6

Leibniz: Unattainable Heights

… one of the supreme intellects of all time.
—Bertrand Russell

First Steps as a Young Man

The British philosopher Alfred North Whitehead studied mathematics at Cambridge University and stayed on there as a lecturer for 30 years before moving to Imperial College, University of London, where he was a professor of applied mathematics for 10 years. During this time, Whitehead was involved in a wide range of fields, including philosophy, producing such a bountiful harvest of work that he was immediately hired as a professor of philosophy at Harvard University upon his retirement and began another distinguished academic career, which he did not leave until he was 76 years old. He died 10 years later in Boston. His early magnum opus is *Principia Mathematica*, which he wrote in collaboration with his younger colleague and former student Bertrand Russell and published in three volumes between 1910 and 1913; the masterpiece of his later career is *Science and the Modern*

Statue of Leibniz (photograph by the author at Leipzig University)

World, published in 1925. In this nearly all-encompassing treatise on natural philosophy, Whitehead referred to the seventeenth century as "the century of genius," which title he also gave to its third chapter.

It is probably the allure of this term "the century of genius" that has prompted me to write three scientific essays in the past 5 years or so: *Fermat's Last Theorem* (in *Southern Weekend*, Sunday, 26 October 2001), *Newton in His Atypical Period* (in *Book City*, No. 6, 2003), and *The Reclusive Frenchman: Pascal and Descartes* (in *Readings*, No. 5, 2006). In other words, I have already written about four scientific geniuses of the seventeenth century—three

6 Leibniz: Unattainable Heights

Frenchmen and one Englishman—and now it is necessary to turn to a German, Gottfried Wilhelm Leibniz: the most erudite man of that century, regarded by Russell himself as the "one of the supreme intellects of all time" (*A History of Western Philosophy*).

Leibniz was born on July 1, 1646, in the city of Leipzig in eastern Germany, a city close to the Slavic Sorbian region, which extends into Poland, leading some recent historians to conjecture that his family may have had Sorbia ancestry; his father was a professor of ethics at the University of Leipzig. His mother, herself a professor's daughter, was his father's third wife, and in fact his father was already nearly 50 when Leibniz was born. The family was more a prototypical bookish family than that of other talented people of his generation. The elder Leibniz personally nurtured his young son during the brief period in which their lives overlapped, so much so that by the age of 8 he was eagerly reading the various Latin works left to him by his father, who by that time had already passed away. By the age of 15, Leibniz was studying law at the University of Leipzig, and at the age of 20 he submitted an excellent doctoral thesis, which was rejected, most likely due to his youth (or because he was too learned, according to Hegel). His mother had died earlier, and Leibniz left his homeland forever. At the beginning of the next year, the University of Altdorf (near Nuremberg) awarded him a doctorate, but he did not accept the offer of a professorship there, in order to get to know the world better. Nor did Leibniz accept an official offer from any university after that; but it would certainly not be accurate to say that he was more interested in politics than in academics.

Leibniz is said to have developed an interest in mathematics while studying Euclid's *Elements* during his time at university. But like the three Frenchmen mentioned earlier, Leibniz pursued his mathematical research only in his spare time. The reason for this was that the university in the seventeenth century was a mere appendage of the Church and philosophy a slave of theology. At the same time, as Russell observed, "most mathematicians were under the shadow of Aristotle's scholastic philosophy, and the impetus for the development of mathematics came from the Renaissance humanists who stood in opposition to the academy." If we compare Descartes with Leibniz, we find that both of them were in love with travel, the former as a soldier and the latter in political and administrative roles. Descartes developed the idea of analytic geometry while stationed in a foreign country, just as Leibniz invented calculus on a diplomatic mission; and they were both under the age of 30 at the time of these remarkable achievements.

Stepping back a bit, we remark that when he was 19, Leibniz had written his first book, *De Arte Combinatoria* (*On the Combinatorial Art*), which he submitted the following year as his habilitation thesis in order to qualify as a lecturer in philosophy at the University of Leipzig and which more importantly established him as a pioneer and even founding figure in modern logic. The main idea of this paper is the formation of propositions as combinations of simple concepts, through which Leibniz hoped to establish an alphabet of thought capable of expressing any human idea. Most importantly, this method was subsequently applied to the interpretation of truth. Leibniz started from the position that all propositions are in the form of subject and predicate (as, e.g., a sentence such as "leaves are green"), a position he maintained throughout his life and developed continuously, as we will see later in our discussion of logic. In light of this, whereas Newton invented a calculus of continuous quantities, which Leibniz also accomplished independently and by his own methods, Leibniz also opened a branch of mathematics in the direction of discrete quantities, known as combinatorial analysis or combinatorics, although this idea did not really become important until the nineteenth and even the twentieth century.

Mathematicians of the Paris Period

Like other great polymaths, Leibniz devoted his youth to mathematics. Somewhat surprisingly, however, the initial passion of such a genius for mathematics was prompted by a political ambition. At the time of Leibniz's birth, Europe had just gone through the 30 years' war, a devastating period of religious conflicts and national movements, which started in Bohemia, but in which the worst losses were suffered by Spain and Germany, especially the latter, which lost most of its population and land after being ravaged by its neighbors. On the other hand, the many local princes who survived were strengthened and largely freed from the rule of the Holy Roman Emperor, gaining de facto sovereignty. In those days, Germany was like China during the spring and autumn and Warring States periods more than two millennia ago, when every lord had a prime minister, ministers, and a group of strategists under him.

Around summer of the second year after obtaining his doctoral degree, Leibniz met with the dismissed chief minister of the Elector of Mainz on one of his journeys. Mainz is most famous today as the birthplace of Johannes Gutenberg, inventor of the movable type press; at that time the Electorate of

6 Leibniz: Unattainable Heights

Mainz was perhaps the most prestigious state of the Holy Roman Empire, and its Archbishop-Elector was responsible for the election of the emperor. The former chief minister was a man of learning and enlightenment, and still very influential in spite of the fact that he had left office, and he was impressed by his encounter with this erudite and humorous young man. He induced Leibniz to accompany him to Frankfurt am Main, then a suburb of Mainz (the relationship between the two places is now reversed). By that time, France had become a major force in Europe, and the Sun King Louis XIV amassed so much power that he seemed likely to invade his northern neighbor at any moment. In view of this, Leibniz, in his capacity as assistant to the legal adviser of the Elector, saw an opportunity to put forward a clever plan of his own above and beyond his official duties helping to redraft the legal code of the Electorate.

His brilliant idea was to divert the attention of Louis XIV from the north with a proposal that would lead to the French conquest of first Egypt and eventually the Dutch East Indies. So, at the age of 26, Leibniz was sent to Paris, where he spent 4 years. Although Descartes, Pascal, and Fermat were all dead by that time, Leibniz was lucky enough to meet the mathematician Christiaan Huygens from Holland (whose father was a diplomat and who was staying in Paris on an annual salary from Louis XIV), responsible for the invention of the pendulum clock and the wave theory of light, among his other contributions. Leibniz quickly came to realize the limitations of the education that he had received in Germany, at that time still in the rear technologically, and embraced his studies under the careful guidance of Huygens with an open mind, especially in mathematics. Due to his diligence and talent, and because the mathematics of the era still had on offer so much low hanging fruit, Leibniz had already made major mathematical discoveries by the time he left Paris (his stated reason for going there having been put aside).

Leibniz's first important mathematical contribution was the binary system, in which he used the numeral 0 to represent the null places and the number 1 to represent the positive places. He himself later observed that the Chinese had hidden this mystery in the 64 hexagrams of the I Ching some three millennia ago. Simultaneously, Leibniz also worked on the development of mechanical calculators, improving on a device invented by Pascal known as Pascal's calculator, which was capable of addition and subtraction, so that it could also carry out automatic multiplication, division, and the taking of roots; at this time the general public was still not so capable of multiplication. He took one of these devices with him to London, another went to the collection of the Hanover Library, and a third was used as a gift from Peter the Great of Russia to the Emperor of China (the whereabouts of this gift today

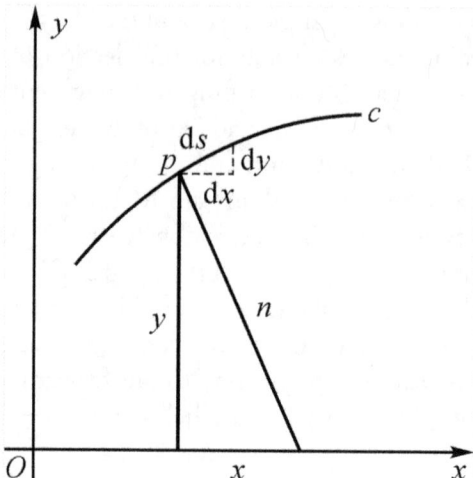

Principles of Differential Calculus by Leibniz

seem to be unknown). It is worth noting that Leibniz did not use the binary system he created for these calculators.

Leibniz's greatest contribution to mathematics was undoubtedly his work with infinitesimals, that is, the invention of calculus. This was an epochal contribution in the history of science, and it was, thanks to this invention, that mathematics began to play an outsize role in the natural sciences and in social life, as well as eventually providing thousands of jobs to people who enjoy mathematics, as happened again the advent of electronic computers in the twentieth century. To his chagrin, Leibniz had to share this honor with Newton, working on the other side of the English Channel. In fact, they both invented calculus independently and around the same time, using different methods (Newton may have hit upon it earlier, but Leibniz was the first to publish his work). Newton's method of fluxions built upon his background in kinematics, and its derivation was more geometrical in character, whereas Leibniz was inspired by Pascal's characteristic triangle and his arguments relied upon more algebraic techniques.

It was the use of algebraic methods, together with Leibniz's own superhuman intuition for mathematical forms (which also greatly benefited his philosophical studies, in contrast with Newton, whose major achievements in the second half of his life were devoted to biblical and theological investigations), that led to his formulation and notational system for the calculus that we know and use today. In addition to this work, Leibniz introduced a

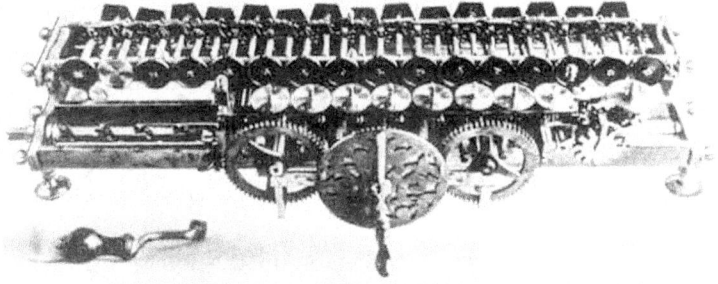

Mechanical multiplication calculator invented by Leibniz

beautifully formed theory of determinants and extended the binomial theorem to any number of variables. Perhaps the most aesthetic of his results is an infinite series presentation for π that he discovered during a trip from Paris to London:

$$\frac{\pi}{4} = 1 - \frac{1}{3} + \frac{1}{5} - \frac{1}{7} + \cdots,$$

known today as the Madhava-Gregory-Leibniz series after the various mathematicians who worked it out across the centuries. The discovery of explicit formulas such as this one put an end forever to the ancient competition for the most precise calculation of the value of π (in which competition Zu Chongzhi had held the lead over the West for 11 centuries).

Logic and Metaphysics

During his time in Paris, Leibniz did not abandon his research into new systems of philosophy, in addition to his concentrated efforts in the domain of mathematics. He managed to get access to the unpublished works of two of his French predecessors, Blaise Pascal and René Descartes, copying them out by hand to review at his leisure. Apparently the publication in Amsterdam of Descartes's *Rules for the Direction of the Mind*, half a century after the death of its author, was based on this handwritten copy. But Leibniz did not become a follower of Descartes, and in fact he struck a position as an anti-Cartesian, especially in physics. Similarly, although he rose to fame in Paris, he never developed any allegiance to France on the basis of his love for the people of his home country, over whom the threat of French invasion hung throughout his lifetime. In addition to his attempt to draw Louis XIV into an expedition to Egypt, he endeavored to weaken the French economy by undercutting the sale of French brandy with cheap rum made from sugar from the West Indies, for example, from Cuba.

By way of contrast, Leibniz always maintained a soft spot for the British, although he stayed at odds with the British academic community because of the priority dispute over the invention of calculus. Leibniz appreciated the assertion of Thomas Hobbes, a British philosopher who lived in Paris for a time, that all reasoning is computation, which may have been a motivation for the invention of his mechanical calculator. Likewise, this assertion drove much of his work in logic. Logic comprises the symbolic study of systems of human thought, incorporating the joint insights of both mathematicians and philosophers. The basic principles of ancient logic come from Aristotle, including the syllogism and the theory of transpositions, but in a direct rather than deductive form. Leibniz sought instead a universal language on the basis of an alphabet of ideas, an abstract propositional calculus and a general methodology. Moreover, he was able to account completely for the syllogisms of Aristotelian logic within his mathematical framework.

Leibniz drew attention to the distinction between the intension and extension and emphasized the independence of intension; for instinct, a proposition such as *all unicorns have horns* counts as true according to this distinction even subject to the observation that there are in fact no unicorns. More significantly, Leibniz introduced a purely formal system of logical deduction. In a series of unpublished logical sketches and a paper entitled *Non inelegans specimen demonstrandi in abstractis*, Leibniz presented a system known as the plus-minus-calculus, along with 24 propositions, including some results that are familiar from modern logic and set theory. For example, if A is in B and B is in C, then A is in C; if $A = B$ and $B \neq C$, then $A \neq C$; $A \oplus B \neq A + B$; and so on. In addition to this, he points out that certain elements of algebra admit an interpretation beyond the remit of arithmetic. This vision of the mathematization of logic was realized two centuries later by the English logician George Boole, who established an algebra of logic, today known as Boolean algebra, which in turn drew back upon the binary system invented by Leibniz. In the twentieth century, another British logician made fundamental contributions in this area: Alan Turing, sometimes referred to as the father of the electronic computer.

6 Leibniz: Unattainable Heights

Leibniz's former residence (photograph by the author, Hanover)

Following upon his work in logic, the second philosophical object of study to which Leibniz devoted himself was metaphysics, which Immanuel Kant called the queen of all sciences and philosophy. Without attempting to nail down a definition, the following statements are generally accepted: metaphysics is the search for the nature of being, and metaphysics is the study of the world as a whole. Etymologically, the word means *after physics* or *beyond physics*; it was coined by one of Aristotle's students in the course of editing his teacher's thought for dissemination. This word has also made an appearance in the disciplines of painting and poetry to designate the so-called metaphysical school of painters and the metaphysical poets, respectively. Leibniz did his work in metaphysics starting from the age of 40, when he experienced a sudden clarity of thought, distinguishing not only between necessary and contingent truths, but also introducing the principle of sufficient reason for truth and the principle of identity of entities. His system of metaphysical thought included, in addition to logic, ideas from linguistics, physics, biology, and physiology and the interconnections among them.

Due to the sudden deaths in quick succession of the Elector of Mainz and the former chief minister, Leibniz found himself without a source of income and was obliged to leave Paris. At the invitation of the Duke Frederick of Lower Saxony, he went north to Hanover as legal advisor and librarian, as well as to write a family history for the Duke. That year he put forward a criticism of the Cartesian description of motion, that is, mechanics, and introduced in its place a new formulation, which has since come to be known as dynamics. Taken together with his refutations of atomic theory and Newton's theory of space-time, among others, his contributions established him as a theoretical physicist at the forefront of his time. A few years later, he improved his system of binary arithmetic and introduced a new discipline called *analysis situs*, which eventually became an important tool in later non-Euclidean geometry and finally transformed into the modern field of topology. In linguistics, as mentioned earlier, Leibniz established that all propositions are in subject-predicate form; in addition to this, he gave what is known as *Leibniz's law*, which states that identical expressions are interchangeable. This, of course, was an outgrowth of his work with the propositional calculus.

Returning to metaphysics, Leibniz argued that the universe consists of innumerable windowless monads, which resemble the soul in varying degrees, and that such monads are unique, simple, and without extension but possessed of qualities and that these monads are the basis for all being. This theory was laid in a celebrated work entitled *Monadology*. Among its consequences is the conclusion that human beings are distinguished from other animals only by degrees, as are living beings from inanimate matter. Descartes had argued that the greatest difference between humans and other animals is that only humans possess consciousness and reason. Leibniz did not go so far as to disagree with this altogether, but he did point out that what triggers our behavior is usually the subconscious, making us more similar to other animals than we might like to think. He also believed in the existence of a subconscious state of mind and that any perception is composed of many microperceptions outside the reach of conscious awareness. Leibniz believed in the interconnection of all things, writing that "any single entity is connected to all other entities"; at the same time, he noted that "every entity is a world unto itself and does not depend on anything other than God."

It should be clear by now that Leibniz was a polymath of the first order. In addition to his work in mathematics, logic, physics, and linguistics, his broad talents also exerted an influence on various fields including geology, botany, jurisprudence, history, and theology, and he even carried out a profound study into the history and religion of ancient China; indeed, he was perhaps the first great Western thinker who was really interested in Chinese culture (followed in short order by Voltaire). Leibniz believed that the ancient Chinese traditions of ritual etiquette (礼), Taoism (道), Tai chi (太极), and so on suggested an alignment with the spiritual force governing the universe, and he was interested in the synthesis of Chinese and Christian metaphysics expounded by the Italian missionary Matteo Ricci, which he defended in an article (without, however, pointing out that this traditional metaphysics and Confucianism lacked a rigorous logical system). Coincidentally, after logic and metaphysics, the third goal to which Leibniz devoted his life was the unification of the two rival religions to which he was committed, Lutheranism and Catholicism, an effort that was unfortunately doomed to be in vain.

The Emergence of the German Nation

The fact that science in England had remained as backward as in Germany until the second half of the seventeenth century is illustrated by the fact that Leibniz was eagerly recruited as a foreign member of the Royal Society in 1673 for bringing a paper and a mechanical calculator to London during a trip of less than 3 months, whereas although Leibniz remained in Paris for 4 years and made major mathematical discoveries there, it was not until 1700 that the Paris Academy of Sciences elected him as a foreign member (this was not a consequence of his hostility to France; Newton was elected that year as well). It was probably also because of this backwardness that the infamous priority dispute between Leibniz and Newton for the invention of calculus developed with such ferocity. The French were always on the side of Leibniz in this dispute, which established his reputation in Europe, and he ended up being censured in England (where the British mathematical community had been cut off from foreign academic exchanges for more than a century).

Before Leibniz, there had been four especially notable German intellectuals: Johannes Gutenberg, the fifteenth-century inventor of movable type, the

Albrecht Dürer self-portrait

printmaker Albrecht Dürer, Martin Luther, and Johannes Kepler, the sixteenth-century astronomer. The first three comprised a technological innovator, an artist, and a religious reformer, respectively, with Dürer considered to be the most mathematically gifted of the Renaissance artists. On the other hand, Kepler did purely scientific research, and he did not have much influence in the humanities or in the broader discourse of ideas. In fact, Kepler's influence was very limited both during his life and after his death due to his long sojourn abroad and his lack of personal charisma. In addition, his personal life was extremely unfortunate, as his first wife and favorite son died of mental illness and smallpox, respectively; the story of his second marriage is an even greater tragedy. At the time of his death, he is said to have been on his way to collect his outstanding payments from his customers. In light of these observations, we conclude that it was really Leibniz who inaugurated modern German science and philosophy and that his great achievements and immeasurable influence lent intellectual confidence to the late blooming of Germany.

Here I cannot help but pause to reflect upon the many more than four giants of culture and thought that had appeared already by this time in China,

Johannes Gutenberg, inventor

among them Confucius and Laozi, who subsequently gained the widest admiration as thinkers and philosophers; Qin Shihuang, who unified China in a way that Germany was long unable to do; Kublai Khan, certainly a figure in Chinese history as Emperor of the Yuan Dynasty, the fact that his grandfather was a great figure in Mongolian history notwithstanding, just as Alexander of Macedon is a figure of Greek history; Cai Lun, whose papermaking technology marks a moment in world history unequivocally as important as printing and the laws of planetary motion; and Li Bai, who possessed a Dionysian temperament and artistic achievements easily rival those of Dürer. But although some of these figures have achieved a historical status or popularity surpassing that of Leibniz, their singular achievements stand as monuments only in one direction or another, and do not stand as a guiding light for the trajectory of a people. Leibniz represents an unattainable height among all nations, looking to the past, the present, the future.

Unusually among the major civilizations of the world, the origins of the Germanic peoples are unknown, and their exact history begins with the conquests of the Romans in the last half century before the Common Era. Even after the sixteenth century, the Germanic peoples were still scattered, divided and in disarray. Although several of the Roman emperors were of Germanic ancestry, their interests were typically internationally aligned, a predilection that was emphasized, for example, in their marriages or recreational pursuits. An example of this trend is Charles V, Holy Roman Emperor, at one point regarded as the greatest monarch in Europe, who always regarded himself as closer to France or Spain in his heart. In the wake of Martin Luther, the northerners for the most part embraced the Reformation, while the southerners wavered between Protestantism and Catholicism. It was when Leibniz had about reached middle age that the Prussian state began to emerge, and it was only in 1701, when Frederick I became King of Prussia and set up his capital in Berlin, that the Germanic nation and identity began to take shape, although still a long way from national unity. Some 40 years later, his grandson Frederick the Great took the throne and greatly expanded the territory.

A polymath responsible for remarkable achievements in science, philosophy, diplomacy, and social activity, Leibniz naturally became the founder and first president of the Berlin Academy of Sciences; the Saint Petersburg Academy of Sciences and the Vienna Academy of Sciences were also established at his initiation, and he is alleged to have even written the Chinese emperor through the mediation of missionaries to suggest the establishment of the Beijing Academy of Sciences; although Kangxi is regarded as the most mathematically minded among the emperors, he did not follow through on this suggestion. Leibniz felt deeply that scholars working independently were expending their energy for small rewards and strongly advocated bringing together talents from all walks of life. He seems also to have exerted influence through his pupil, the daughter of Duke Augustus, on her husband, the future Frederick I, in order to establish the Berlin Academy of Sciences. In any case, the Berlin Academy of Sciences quickly became one of the four most influential research institutions in Europe and attracted two of the most outstanding scientists of the eighteenth century, Leonhard Euler and Joseph-Louis Lagrange. Prior to this, Leibniz also supported the founding of the *Acta Eruditorum*, an influential Latin journal in the history of modern science.

It is true that the rise of Germany depended on the strength of the Kingdom of Prussia. But in the 72 years following Leibniz's death, Germany was home to a succession of great philosophers—Kant, Fichte, Hegel, Schelling, and Schopenhauer—transforming the world of German thought into a star-studded one. Hegel incidentally was born in the same year as Beethoven, in which year also Goethe failed to obtain his law degree at the University of Leipzig and transferred to the University of Strasbourg. Of these, Kant is often considered the greatest philosopher of modern times, a pure man who dedicated his entire life to his intellectual pursuits. But Kant's philosophy was in many ways influenced by fellow philosopher Christian Wolff, himself a disciple of Leibniz, and the ideas of the master and the apprentice constitute a complete philosophical system. In contrast, Germany was slightly slower to flourish in the exact sciences, but after Gauss, the prince of mathematics, born 7 years after Hegel, reached maturity, the world center of mathematics also shifted from France to Germany, from Paris to Göttingen; Gauss himself, incidentally, acknowledged Leibniz as the highest mathematical intellect. From that time until now, except for the interruption of two world wars, Germany has been a power at the forefront of world civilization.

A Very Busy Stranger

In the era in which Leibniz lived, he was seen by the public as a typical late Renaissance humanist. He himself was an optimist, believing that the world in which we live is the best of all possible worlds. Even his talents, however, had limits: his lifelong aspirations to success on the literary stage, and the pride he took in the poetry he wrote, mostly in Latin, could amount to no more than wishful thinking. More seriously, Leibniz did produce any complete and celebrated scholarly monograph, as Descartes, Pascal, and Newton did, but rather disseminated his thoughts only in fragments, in the form of notes, letters, and brief articles. This is partly because he was an amateur scholar, devoting only his leisure time to research, and partly because of what Russell has described as his philosophical duality, often using metaphysics and logic to express the same thought, even though the works on logic were not published until two centuries after his death.

As a scientific and philosophical star who sprung up overnight in a nation still behind in its development on the intellectual scene, Leibniz inevitably developed some unfortunate habits, the most prominent of which was a love of vanity. At one point in his later years, he was employed by five royal courts—Hanover-Brunswick, Nuremberg, Berlin, Vienna, and Petersburg—at the same time. And he was perpetually promoting amazing initiatives far beyond his ability to realize them. For example, he believed that the German economy might be revived by introducing the production of silk fabrics, and for this purpose he personally planted Italian mulberry seeds in his yard; he proposed the establishment of a public health system in Berlin, and a system of fire services; at the same time, he directed the design of the gardens of the royal palace, proposed the construction of streetlights in Vienna, a national bank, isolation wards for plague victims, and a plan for the management of the Danube river;

he initiated research projects on the existence of an isthmus between Russia and America, the origins of the Slavs and their language, and so forth.

Perhaps it was due to the influence of metaphysics and because he devoted so much of his time and energy to worldly matters, including traversing the rugged mountain roads of Europe in a rickety four-wheeled carriage, that Leibniz remained unmarried for the duration of his life, despite maintaining close relationships with many royal women and even serving as an intermediary between the princes and princesses and declaring himself an expert on the topic of succession to the throne. He had this in common with the ancient sages he admired—Thales, Heraclitus, Plato, Omar Khayyam—with his close contemporaries and intellectual rivals such as Descartes, Pascal, Baruch Spinoza, and Newton and his German intellectual successors, Kant, Schopenhauer, and Nietzsche. Perhaps it was the case that celibacy better suited a figure so committed to the life of the mind. But I surmise that his proud heart must have been hurt at some point by some princess or noblewoman. Later, near the age of 50, Leibniz described his own life as full of confusion. Biographers describe him as a man who did not give up once he had set his sights on a goal and who corresponded with hundreds of people. These letters show that he was an intellectual with an impetuous temperament but who expressed his ideas as quickly as possible.

On November 14, 1716, an unremarkable day, Leibniz died in the presence of his secretary and coachman after having been bedridden for a week with colic caused by gout and gallstones. By this time, his second employer, Duke John Frederick of Brunswick, and his brother Ernest Augustus, whose wife Sophia of Hanover was an admirer of Leibniz, had already passed away, and Augustus's eldest son, Georg Louis, had married into the British throne 2 years earlier. But due to the ongoing priority dispute over calculus and his own old age, he was not able to secure an invitation to London such as when the Elector of Mainz had visited, and his sense of loneliness grew. Nor had he found favor in Hanover, where he was resented by the local senators and his parliamentary colleagues as a foreigner and for his fashionable, elegant style and frequent international travels, so much so that they refused even to attend his funeral. Leibniz was buried in an unremarkable cemetery, and, according to E.T. Bell, only his secretary and the gravediggers heard the sound of dirt falling on his coffin.

Nearly a century later, Napoleon's armies invaded Germany, and the French discovered a large number of manuscripts left behind by Leibniz in the Royal Library of Hanover, including a brilliant plan for the conquest of Egypt, which was to be presented to Louis XIV. This ambitious plan had already been implemented a few years earlier by Napoleon, who is said to have been dismayed to learn that Leibniz had thought of it before him. As the latter had

6 Leibniz: Unattainable Heights

Leibniz's tomb (photograph by the author, Hanover)

expected, the military campaign, which was intended to threaten the road to India and cut off the fortune of Great Britain, ended in failure after a brief victory. Moreover, the military defeat in Egypt directly led to the total defeat of the French army in the Apennines. On the other hand, several of Napoleon's men stumbled upon a stone inscription in three scripts at Rosetta near Alexandria, which helped later archaeologists to solve the mystery of ancient Egyptian hieroglyphics and thus revealed the secrets of the ancient Egyptian civilization, including their knowledge of geometry.

January 2007, Hangzhou

7

John von Neumann, Who Made the World a Better Place

I recognize the lion by his claw.
—*Johann Bernoulli*

An Incredible Mind

He was the son of a wealthy banker in Eastern Europe, a bohemian who liked to go out to nightclubs and yet became a pivotal figure in the twentieth century; before the onset of the Second World War, he was an outstanding mathematician and physicist, the youngest among the first five tenured professors hired by the Institute for Advanced Study at Princeton (he was 29 at the time, while the eldest of them, none other than Albert Einstein, was 54). During the Second World War, he proved indispensable to the Allied forces, including the Army, the Navy, the United States, and the United Kingdom, because he was the foremost expert in explosion theory and the designer and promoter of the first atomic bomb. After the war, the field of game theory, which he founded, greatly expanded the research of mathematical economics and influenced the work of at least 11 Nobel Prize in Economics recipients. His greatest contributions may have been in the theory and practice of computer science, and he is considered a founding father of the electronic computer. In short, he was the most useful talent to come from abroad to the United States in the twentieth century. This is the subject of this chapter, none other than John von Neumann, who was born in Hungary into a Jewish family not long after the turn of the century.

Von Neumann in his study

Von Neumann was stocky, with bright brown eyes and a face accustomed to smiling; such features are commonplace enough, but to achieve such rich and significant achievements as he did, one must have an unusual mind. For one thing, he had an astonishing memory for the things he took an interest in and was able to recite at will full pages from the novel *A Tale of Two Cities* by Charles Dickens or noteworthy entries from the *Encyclopedia Britannica* that he had read some 15 years earlier. As for mathematical constants and formulas, he had a brain full of them and was able to bring them to the surface at any time. His reading speed and ability for calculation were no less amazing. As a boy, he is said to have sometimes taken two books with him when using the restroom. Later, after he had already become famous, his assistants and graduate students often felt as though they were riding a bicycle trying to keep up with the speeding train carrying von Neumann along. During calculations, he typically took a strange look, staring up at the ceiling with an expressionless face. During this time, his mind was running at a high speed. If he happened to be on a high-speed train, his thoughts and calculations would also speed up accordingly.

If the various abilities mentioned above convey an almost supernatural feel, the following aspect of von Neumann's excellence is not unattainable to mortals: this is the desire and commitment to constantly learn new things. During one summer vacation from his time as an undergraduate student in the Department of Chemistry at the University of Berlin, von Neumann returned to his home in Budapest, where he met a fellow student who was preparing to attend Cambridge to study economics; immediately interested, von Neumann asked him to recommend some introductory texts in economics and from that time on sustained an interest in this new subject. On another later occasion, he was invited to London to train the British Navy in the detonation of German mines; but while he was there, he studied aerodynamics and developed a strong interest in computing technology. The former pursuit saw him developed into a pioneer in the study of oblique shock waves, while the latter led him to become a leading researcher in numerical analysis. His direct involvement in electronic computers came about through an encounter on a railway platform, and in general von Neumann was always especially prolific during his travels. The remarkable trend in each of these cases is that almost all of his achievements were carried out during a time when he was primarily engaged in some other pursuit.

As a person who was compelled often to collaborate not only with leading scientists, but also sometimes with politicians and military strategists, who served in capacities ranging from president of the American Mathematical Society to special advisor to the President of the United States, von Neumann needed also a keen political sense and poise. Prior to the Second World War, von Neumann predicted that Germany would conquer a weak France and that the Jews would suffer genocide similarly to the Armenians in Turkey during the First World War, after which the United States would benefit from conflict between two powerful enemies (Germany and the Soviet Union). He concluded moreover that the Soviets would arrive at nuclear weaponry sooner or later, since the secret of the atomic bomb was a simple thing and anyone with an education could develop it. As for his predilection for equilibrium, this may have been something preternatural rather than driven by ambition; in any case, he had no need to spend money to improve his public relations. A further remarkable characteristic: von Neumann was not ostentatious, and did not like to argue with others. When faced with tense situation, he was prone to diffuse it with lighthearted jokes and anecdotes.

Of course, von Neumann's ingenious mind was not without its shortcomings; most notably, he was not as original a thinker as Einstein or Newton. But he was able to quickly seize upon the innovative sparks and concepts of others and develop them in depth in detail into fully realized theories for use by the academic community and mankind in general. By comparison, after Einstein immigrated to the United States, his role as a thinker was in the main symbolic, even decorative, whereas what von Neumann achieved was irreplaceable. Admiral Lewis Strauss, a member of the nuclear hawks, said of von

Neumann that he had an inestimable ability to get to the heart of a problem, to break it down, until suddenly the most difficult questions became simple and clear and everyone was left wondering why they had not been able to see the problem so clearly and arrive at an answer. Eugene Wigner, recipient of the Nobel Prize in Physics, responded to a question about the influence of von Neumann on United States government's formulation of scientific and nuclear policies that once Dr. von Neumann had analyzed a problem, it became clear what to do.

When so many qualities are added up together and concentrated in a single person, his advantages stick out quite prominently. Wigner had grown up with von Neumann in Budapest and admitted that he developed an inferiority complex with respect to this high school colleague, 1 year his junior. After winning the Nobel Prize, he was interviewed by Thomas Kuhn, the famous historian of science and author of *The Structure of Scientific Revolutions*, who remarked, "… you have a good memory," to which Wigner replied "Not like Neumann's." Kuhn observed that it "must have been a shattering experience– to have grown up with Neumann, however bright one is."[1] Yet another recipient of the Nobel Prize, the German-American physicist Hans Bethe, who like Wigner was a colleague of von Neumann's from the Los Alamos Laboratory days, once remarked, "I have sometimes wondered whether a brain like von Neumann's does not indicate a species superior to that of man." He was one of very few people in human history who could change the course of the world by writing a few formulas on a blackboard. The French mathematician and Bourbaki member Jean Dieudonné believed that von Neumann was the last of the great mathematicians.

Family Gatherings over Lunch

John von Neumann was born in Budapest, the capital of Hungary, on the banks of the Danube River, on December 28, 1903, and given then name Neumann János Lajos. Although Hungary and Austria at that time were united as the Austro-Hungarian Empire, the nature of their alliance was purely diplomatic and military, while the two remained independent in their domestic affairs and economies and preserved their own separate country names, kings, and languages. Unlike the vast majority of Europeans, the

---

[1] Interview of Eugene Wigner by Thomas S. Kuhn on 1963 November 21, Niels Bohr Library & Archives, American Institute of Physics, College Park, MD USA, www.aip.org/history-programs/niels-bohr-library/oral-histories/4963-1

7 John von Neumann, Who Made the World a Better Place

Hungarians, like the Chinese, place their surnames first, followed by given names. This has been an important foundation for the conclusion of scholars that their ancestors came from Central Asia or the Mongolian steppes. The name János is equivalent to the English name John; and they take, respectively, the similar nicknames Jani and Johnny. When von Neumann was 10 years old, his father, a banker, was awarded the title of noble for meritorious service as an economic advisor to the government, from which point on the honorific *von* appeared in front of the family name to make *von Neumann*, so that, after he immigrated to the United States, he went by John von Neumann.

For 35 years before von Neumann was born, Budapest had been the fastest growing city in Europe, just as New York and Chicago (victors in the Civil War) were the fastest growing cities in the Americas. The population jumped from 17th place to 6th, following London, Paris, Berlin, Vienna, and St. Petersburg. Budapest was also the first to go electric, installing the first electric subway in Europe and replacing stagecoaches with trams, eradicating the problem of disease spread through horse manure. The Elisabeth Bridge across the Danube was built in the year of von Neumann's birth, the longest suspension bridge in the world at that time. This was the golden age of Hungary, and Budapest had something of the mood and atmosphere of Paris, with more than 600 cafes, an opera house with acoustics surpassing the opera house of Vienna, caretakers from around the world to look after the children, and nightclubs famous for charming girls patiently attending to the political opinions of their guests.

In the half century prior to the outbreak of the First World War, Budapest and New York City were the most popular destinations for immigration for educated Jews from around the world. In these two urban paradises, they quickly attained professional success as doctors, lawyers, or businessmen. Of these, the Jews who immigrated to New York City were of humbler origin, an artifact of the limitations of transportation at that time: tickets for lower class steerage cabins across the Atlantic were relatively cheap. Only a few of the very wealthy could afford to travel by luxury liner, and in any case their livelihoods across the ocean were by no means guaranteed. So Budapest was the more desirable destination among middle-class and upper-class Jews, not least on account of its excellence in secondary education. Of particular importance, Jews in other Central European countries still occupied at that time an inferior position, whereas something had changed in Hungary, primarily because the Hungarian Jewish population had maintained solidarity with the Magyars, the main ethnic group in Hungary, at a time when various ethnic minorities had taken to rioting. They were repaid for this with the repeal one by one of

Von Neumann at six years old

discriminatory laws. Among the Jews born in Budapest was Theodor Herzl, father of modern political Zionism.

Moving further back in time, von Neumann's ancestors hailed from Russia. His father had been born in a small town in southern Hungary, not far from Serbia, where he received a solid rural education. He made his way to the capital after graduating from middle school, passed the bar examination, and started a successful career with a bank. Among his wide circle of friends was a doctor of law who later became a Lord Chancellor of the Court of Appeal; eventually the two of them became brothers-in-law, bringing both of them through marriage into a prosperous Jewish family. Von Neumann's grandfather had collaborated in agricultural equipment, learning successfully from the sales experience of the American Sears Company. His four daughters all brought sons-in-law into the family, which occupied both sides of a bustling commercial street in Budapest, with shops along the bottom floors and residential units above. Von Neumann grew up above the store, like the British politician Margaret Thatcher, born more than two decades after him.

Up to the age of 10, von Neuman received a typical Jewish education: he was taught by tutors. At that time tutors and nursery maids were viewed as an essential amenity for the middle and upper classes. The learning of foreign languages was especially prized, as many parents believed that children who could speak only Hungarian would encounter problems in their futures. First was German and then French and English; slightly older to children continued on to Latin and Greek. Latin in particular had been taught in Hungary for hundreds of years and was considered an axiomatic language capable of bringing organization and logic to human minds. Perhaps it was his early training in Latin that led von Neumann later to the creation of computer

7 John von Neumann, Who Made the World a Better Place

languages. Of course, mathematics was also essential, and von Neumann exhibited a talent in his childhood for computation, being able to quickly calculate the product of two four-digit or five-digit numbers, something he inherited from his maternal grandfather. Early on, von Neumann observed that mathematics is not abstract and boring but follows certain laws. From his mother's artistic talents, he discerned also the elegance of numbers, a requirement for his later academic pursuits.

Another passion was history. Apparently he once read through a history of the world in 44 volumes, filling its pages with notes on small slips of paper. But he was not omnipotent, proving mediocre in fencing and music, and later even resented being referred to as a professor, as it was the title of his fencing coach. His family had hired an excellent cello teacher, but von Neumann never seemed to escape the stage of practicing fingerings. Still, there were many great conductors and pianists among the Hungarian Jewish population. Among those who made their ways to the United States were Georg Solti, who conducted in Chicago, Eugene Ormandy in Philadelphia, George Szell in Cleveland, and Antal Doráti in Dallas. It was also Hungarian immigrants to the United States who built Hollywood, founding Fox and Paramount. Von Neumann Sr. also began to invest in the film industry and theater after he had achieved success in the banking industry.

Here it must be mentioned that the von Neumann family had a great tradition of family gatherings at lunchtime, where the children rush to propose questions for everyone to discuss, for example, a certain poem by Heinrich Heine, the dangers of antisemitism, the sinking of the Titanic, the achievements of their grandfather, and so forth (contrast this with the cafeterias at Chinese universities today, where the most indispensable feature is a high-definition television). In those days, adults did not impose their views upon children. At one of these gathers, von Neumann proposed that the retinal imaging principle of the eye should be different from the behavior of small particles on photographic negatives; he argued that the input must be in multiple channels or regions, although the ear might have a single channel or linear input. Throughout his life, he sustained an interest in the difference between the operation of the central nervous system and the technology of machines or robots with human inputs. The first time he watched a film with sound he was surprised that although the sound was produced by speakers that were invisible on the screen, it seemed to come from the mouths of the actors. This was a happy childhood, and later von Neumann's two wives were girls who lived next door and with whom he played as a child.

At the age of 10, von Neumann began to attend school at what was generally referred to as a public school or grammar school in English and French and gymnasium in German. Countries that took Germany as their spiritual touchstone, which included Austro-Hungary, also used the word *gymnasium*, which derived from the sporting arenas in which young men in ancient Greece had competed naked. The education system in Hungary at the time was an elite one and fiercely competitive, in which one-tenth of the most capable students were carefully cultivated and the rest left to their own devices. The Jewish community prospered under this system, and it was felt that they were better suited to the study of rational data than to human interactions or as Einstein later put it to escape from the world of people and pass to the world of things. At the end of the Second World War, Japan imitated the education system of Hungary, measuring the standard of a high school by the numbers of its students admitted to the University of Tokyo, which contributed to its rapid economic improvement and produced dozens of Nobel Prize and Fields Medal recipients. The Japanese were trying to catch up to the Americans, who had defeated them, as the Hungarians earlier sought to surpass the Austrians, toward whom they nurtured an antipathy. The education system in China by contrast does not currently operate under any such motivation.

The school at which von Neumann enrolled was a Lutheran one, and the instruction was carried out in German. Around the same time, a total of four Jewish boys matriculated to three of the top schools in Budapest, and all of them later emigrated to the United States. In addition to von Neuman, these were Leo Szilard, Eugene Wigner, and Edward Teller. It was in large part through the contributions of these four Hungarians that the United States successfully developed the atomic and hydrogen bombs. Indeed, the latter three physicists had persuaded Einstein to send his famous letter, which had in fact been drafted by Szilard, to President Roosevelt in the summer of 1939, suggesting the development of the atomic bomb and directly instigating the start of the Manhattan Project. Subsequently it was Zillard who first proposed the concept of a chain reaction, while Wigner established the theory of neutron absorption and assisted Enrico Fermi in the construction of the first nuclear reactor; but it was Teller (who supervised the doctoral work of Chinese physicist Chen-Ning Yang) who became known as the "father of the hydrogen bomb." These four Jewish scientists operated from a place of fear and loathing for Nazi Germany and the former Tsarist Empire, prompting them to devote themselves to the development of nuclear weapons regardless of the risks.

It was also in 1914 that the First World War broke out; the Austro-Hungarian Empire declared war on Serbia in response to the assassination of Archduke Franz Ferdinand of Austria, presumptive heir to the throne, and Russia and Germany were quickly drawn into the fray. The von Neuman family was exempted from military service on account of their high standing and remained able to travel to Venice and other places during the war. Following the victory of the Allied Powers, the last czar Nicholas II of Russia was replaced by the red regime of Vladimir Lenin, and some two-thirds of Hungary's land was carved up by neighboring countries. This did not however have an effect on its elite education. The principal of the middle school at which von Neuman was a student admired his mathematical talents and recommended him to a professor at the University of Budapest. At the age of 17, von Neuman worked in collaboration with a professor of mathematics to study the roots of Chebyshev polynomials and published his first work in a German journal. The year 1921 saw a successful end to his high school career when von Neuman was awarded the Eötvös Prize, a national award for mathematics. Other recipients of this award include Zillard, Teller, and the engineer Theodore von Kármán, the father of supersonic flight, who also happened to be the doctoral advisor of Qian Xuesen, who founded the nuclear industry in China.

Wandering Through Europe

During his final year in high school, von Neumann's father began to worry about his son's future. He consulted many of his friends, including von Kármán, at that time undersecretary for education for the coalition government formed by the Communists and Social-Democrats. Finally, it was determined that von Neumann would study chemical engineering, which was viewed as a promising choice similar to computer science and biology in China at the turn of the new millennium, and the will of the older generation was imposed upon the younger. So von Neumann attended the University of Berlin and the Federal Institute of Technology Zurich (ETH Zurich) as a student of chemical engineering; but really he wanted to study mathematics in spite of the fact that the future of mathematicians in Hungary was not promising; he registered as well as a doctoral candidate in mathematics at the University of Budapest. A young man under the age of 18, von Neumann found himself splitting his time between interdisciplinary undergraduate and graduate studies in three distant cities, indicating the perseverance and intellectual confidence that ran in his family.

Von Neumann arrived in Berlin, the capital of Germany, in autumn of 1921. The original plan was that he would pay a visit to the famous Jewish chemist Fritz Haber, an impressive figure who had invented poison gas in 1915, much to the advantage of Germany, which at that time was besieged on all sides in the course of the First World War, and received the Nobel Prize in Chemistry in 1918, the same year in which Germany was defeated, for his

The main entrance of the old building of the Institute of Mathematics in Göttingen (photograph by the author)

discovery of a method to synthesize ammonia from nitrogen and hydrogen gases. The winner of the Nobel Prize in Physics that year was also the German Max Planck, founder of quantum mechanics. In fact, Haber was the only recipient of a Nobel Prize in Chemistry during the First World War, which speaks to the fact that Sweden did maintain its neutral position throughout the war. But instead von Neumann passed through two unexpected lost years: he did not visit Haber and often missed his chemistry classes. At that time, the sex service industry in Berlin was notoriously licentious and very affordable for young people holding foreign currency. His family back home were well aware of these worrying circumstances.

In fact, their worries were somewhat misplaced. A person such as von Neumann, active and possessed lofty ideals, would not be satisfied with a single professional direction nor indulge overmuch in a single vice or entertainment. During that period, his main interest was set theory. Although he had attended lectures by Einstein in statistical mechanics at the University of Berlin, he was more influenced by mathematics professor Erhard Schmidt, who had studied early on under David Hilbert, and counted among his friends Ernst Zermelo, who had taken the lead in formulating an axiomatic set theory avoiding the problems of Russell's famous paradox. Unfortunately, there had remained an ambiguity in Zermelo's formulation, which was resolved several years later by Abraham Fraenkel, forming what it now known as Zermelo-Fraenkel set theory, or *ZF*. This system was the last hope for a proof of Cantor's continuum hypothesis after the Austrian mathematician Kurt Gödel proved his incompleteness theorem in 1931, although this hope too was dashed in 1963 by the American mathematician Paul Cohen, who received a Fields Medal for this result.

In 1923, von Neumann finally completed a long thesis and submitted it to a German journal of mathematics, at which Schmidt served as editor. The latter passed it along to Fraenkel for review, who was deeply astonished by it and immediately invited von Neumann to Marburg in central western German as a guest, and finally suggested to von Neumann that he publish it under the title *An Axiomatization of Set Theory*. This system was later improved upon by the Swiss mathematician Paul Bernays and Gödel, forming what is known as von Neumann-Bernays-Gödel set theory, or *NBG*, which has been proven to be a conservative extension of *ZF*, and remains one of the firmest foundations for set theory. It is worth noting that von Neumann included toward the end of his paper the observation that no known method was able to avoid all difficulties, suggesting that he had vaguely foreseen something of the revolutionary results of Gödel still to come.

Many years later, after he had immigrated to Israel, Fraenkel reflected on these events and observed that he himself had concluded at the time that this was a tremendous article, going so far as to invoke the famous remark made by the eighteenth-century Swiss mathematician Johann Bernoulli upon reading an anonymous paper by Newton: "I recognize the lion by his claw." The paper itself was soon circulated among the mathematical heavyweights of the era, although it has not yet been published, and from that time on its author, still a young undergraduate chemistry student, was regularly invited to Göttingen, where he was a frequent guest at the home of David Hilbert, the great mathematical master. These two men, one young and one old, with more than 40 years of age separating them, spent hours in the study or the garden, causing some vexation among the professors of Göttingen. In short, the lost 2 years of von Neumann can be compared to the 2 years that Isaac Newton had spent in his hometown avoiding the plague in the seventeenth century, during which period he took advantage of his time to invent modern science. Of course, Newton had been on a quiet farm, and von Neumann in a bustling city. These were the same 2 years in which two masterpieces of modernist literature were published: *The Waste Land* by T.S. Eliot and *Ulysses* by James Joyce.

We turn next to Zurich. During his time in Berlin, von Neumann had taken only some basic courses in chemistry, but he had to travel to Zurich to obtain his degree. In the autumn of 1923, he easily passed the entrance examination of the ETH Zurich, which Einstein had taken twice, and began the second stage of his studies. During his first semester, he received excellent marks in each of his subjects, including organic chemistry, inorganic chemistry, and analytical chemistry. Interestingly, although he was already so accomplished in mathematics, he was obliged to take the most basic courses in advanced mathematics. Two years later, he had just barely completed his coursework in chemical engineering, having broken in the meantime countless glass containers in the laboratory. By that time, he had already established a close relationship with the two best mathematicians at the university, Hermann Weyl and George Pólya. When Weyl had to give lectures or attend conferences elsewhere, von Neumann, despite being an undergraduate in the Department of Chemistry, would take his place in the lecture hall. Pólya, a fellow Hungarian and senior classmate, claimed to have been scared of the questions von Neumann would ask him, recalling a time when he had mentioned an unsolved problem in the course of his lecture only for von Neumann to come to him at the end of the lecture with its solution.

In the summer of 1925, von Neumann received his bachelor's degree in chemical engineering from the ETH Zurich, and the following spring he also delivered the defense of his doctoral dissertation in mathematics at the University of Budapest at the age of 22, under the nominal supervision of Lipót Fejér, head of the Department of Mathematic, who also served as advisor to Pólya, Paul Erdős, and Pál Turán. Hilbert arranged for von Neumann, who was fascinated by quantum mechanics at the time, to come to Göttingen as a visiting scholar, and subsequently he secured employment as a privatdozent in Berlin, an unpaid lecturer position established in Germany for young people intending to pursue an academic career, which also served as the only path to professorship. In addition to its lack of salary, this position brought with it no security; the only remuneration was from the tuition fees of students attending courses, which imposed the requirement of a heavy workload. For von Neumann, who was wealthy, this was no hurdle, but it was undoubtedly torture for someone like Einstein, who came from a poor family, which probably explains his time in Bern working as an obscure patent clerk rather than lecturing at a university.

After 2 years at the University of Berlin, von Neumann transferred to the University of Hamburg, where he established a series of important research results in set theory, algebra, and quantum mechanics while serving as an unpaid lecturer, attracting the attention of the mathematical community. For example, in measure theory, he published a paper *On the General Theory of Measures*, in which he extended the notion of measure from classical Euclidean space to the setting of an arbitrary group, not necessarily abelian, and showed that according to his general definitions every solvable group is measurable. This paper however cannot be said to occupy a central position in its discipline as had his earlier work on the axiomatization of set theory. For another example, in his work on operator theory, von Neumann gave the first abstract definition of a Hilbert space; later he extended the spectral theory of selfadjoint operators on such spaces from the bounded case to the unbounded case, obtaining in the process a spectral theorem of his own, a necessary precondition for the birth of functional analysis, the flower of abstract mathematics. But it was his work on quantum theory that was his best and most important during that period.

Although classical Newtonian mechanics rules the everyday world and applies to everything that can be seen with the naked eye, it was discovered in the early twentieth century that it fails to adequately describe the physics of matter which is either moving too quickly, in which case certain laws of Einstein's theory of relativity come into play, or which is too small, in which case the laws of quantum mechanics take over, enabling the description of the states of molecules, atoms, and electrons, the word *quantum* deriving from Latin for *how much*. Planck had discovered that light, X-rays, and other waves occur only in discrete parts, each of which is called a *quantum*. Quantum mechanics is the very foundation of modern theoretical physics and technology and led directly to the electronic revolution, as well as to the birth of the atomic bomb, and one of its central points is the mathematical description of the atomic state, for which von Neumann provided an entirely new form: the state of an atom is represented by a unit vector in a Hilbert space. It was this formulation that unified the two competing representations of quantum mechanics, namely, the matrix mechanics of Werner Heisenberg and the wave equation of Erwin Schrödinger.

America During the Great Depression

Von Neumann had been willing to accept an unpaid lectureship position at the University of Hamburg because there were more opportunities for advancement there than in Berlin; notably, his father had died that spring. If he had stayed on and actually become a professor at the University of Hamburg, the Chinese mathematician Shiing-Shen Chern might have become his student after a few years. But in the fall of 1929, Princeton University in the United States unexpectedly extended to him an offer to visit as a guest lecturer. It turned out that his teacher Weyl from Zurich was spending a year there as a visiting professor, until the retirement of David Hilbert obliged him to return to Göttingen to take over the position of his teacher, although the two of them to have divergent philosophical views. So it was Weyl who recommended von Neumann to the Americans. Of course, the position of guest lecturer was offered as a transitional role, and due to the scarcity of professorships in Germany and the approaching war, the United States gained a rare talent. Before his journey, von Neumann returned first to Budapest, where he converted to Catholicism, and completed a major milestone in his life: he got married, and later his new wife came with whom to Cherbourg, France, to cross the Atlantic by boat.

Although the United States was deep in the midst of the Great Depression at the time, von Neumann immediately fell in love with this country of immigrants, where people were focused on practical results, stuck to themselves, and did not abide too much by convention. The following year, he successfully obtained his promotion and became a professor. Two years later, he became one of the first tenured professors to be hired by the ambitious newly established Institute for Advanced Study at Princeton, partly due to the temporary withdrawal of Weyl: although von Neumann had originally been considered for the position, this decision had not been finalized, due to his young age and reluctance on the part of Princeton University, and ultimately Weyl had accepted it upon learning that Adolph Hitler had become the chancellor of Germany. Including Einstein, three of the first five professors hired by the Institute for Advanced Study were German, and of these only Weyl was not Jewish, although his father-in-law was. The following statistic is instructive: of the scientists who immigrated to the United States from Germany in the years following 1933, 11 had either received a Nobel Prize or would go on to do so, and 10 participated in the Manhattan Project.

Speaking of Princeton, more specifically the Institute for Advanced Study, it occupies a special place in the academic world, inspiring envy and

7 John von Neumann, Who Made the World a Better Place

Von Neumann and Oppenheimer (second from right)

admiration. Even during the Great Depression, the professors there were paid well to work completely free of obligations. Its structure is also unique and somewhat strange: there are no graduate students, and only the most outstanding doctoral students are recruited to carry out postdoctoral research as a springboard for their future careers. There are physicists, but no laboratories, which many people found objectionable, among them J. Robert Oppenheimer, leader of the Manhattan Project, who refused to become a professor there, although later he served as its director, viewing it as a madhouse where egocentric stars lived alone and desperately, shining their own lights in the darkness. The mathematician Richard Courant, another student of Hilbert, and the physicists Richard Feynman, who received a Nobel Prize, both also felt that teaching and students are necessary components of meaningful academic life and that the relevant questions proposed by students are a source of stimulation during periods of low inspiration.

Einstein himself experienced at times a certain oppressive atmosphere there, writing in a letter to the physicist Max Born in Edinburgh that he felt like a bear hibernating in its den. When Einstein had reached the United States from Europe, the mayor of New York brought a band to wait for him on the pier, but the Institute for Advanced Study arranged for a motorboat to take its guests directly to the New Jersey shore. When President Roosevelt and his wife invited Einstein to dinner, their invitation was politely declined on his behalf. Some people hold to the view that great scientific minds have no need for the fireworks from the social world, that their ideas form in a vacuum. Einstein remained diligent, looking for contradictions in the theory of quantum mechanics, but to no avail. In a letter to Queen Elisabeth of Belgium, Einstein described a strange and rigid village, inhabited by people with reputations difficult to live up to and signed it with the placename *concentration camp, Princeton*.

Even von Neumann was not happy at Princeton, and neither he nor his wife and daughter could wait for the summer to come each year and with it their return to Europe. After the third summer, his wife, who had suffered frequent humiliations at Princeton due to differences in her lifestyle, did not make the return trip with him to the United States. Their only daughter stayed with her, and was not reunited with her father until she was in middle school. But von Neumann never stepped away from his academic research and continued to obtain outstanding results. During the 1990s, as the Institute for Advanced Research was preparing to celebrate its 60th anniversary, they carefully highlighted three landmark achievements carried out there: the research of Kurt Gödel into the continuum hypothesis, the work of von Neumann, and the work of Chen-Ning Yang and Tsung-Dao Lee on parity violation. To this it is now necessary to append the proof of Fermat's last theorem by Andrew Wiles.

During his time teaching at Princeton University, von Neumann summarized the mathematical development of quantum mechanics and published it as a book entitled *Mathematical Foundations of Quantum Mechanics*, which remains to this day a classic work. He also derived the famous weak ergodic theorem in statistics. During the year in which he was appointed to the Institute for Advanced Study, he made a breakthrough in his research in group theory and solved Hilbert's fifth problem for compact sets. Together with a young mathematician named Francis Murray, he wrote a series of papers on operator rings, a natural extension of the finite-dimensional theory of matrix algebras, which later became a powerful tool in quantum mechanics; a by-product of this research was the appearance of continuous geometry. The subjects of this research are referred to today as von Neumann algebras in his honor. In lattice theory, von Neumann was the first to discover the infinite distributivity of the operations of union and intersection in Boolean algebra and its equivalence to continuity. All of this can be attributed to his youthful vigor and unceasing creativity; he was a person who knew how to relax and think.

Mines and the Manhattan Project

In 1937, von Neumann became an American citizen. That same year, Japan launched its full-scale reactionary war of aggression against China, and over the course of the next 2 years, Germany annexed Austria and Czechoslovakia, and Italy annexed Albania. But the start of the First World War is pinned by Western historians to September 1939, when Hitler's troops invaded Poland, and Britain and France declared war on Germany. By this time, preparatory military activity was already underway in Aberdeen, a coastal town in northeastern Maryland, and von Neumann was invited there to serve as a consultant for the Ballistic Research Laboratory under the Office of the Chief Ordnance of the US Army (later the Scientific Advisory Council). This brought into his view a new subject: artillery ballistics. Scientists had long known that resistance and trajectory of artillery shells passing through air with changing concentrations are governed by nonlinear equations, which in general are difficult to solve. Von Neumann became a calculator of shock waves and ballistic trajectories in the precomputer era.

He did not however devote much energy to this project until the Japanese attack on Pearl Harbor at the end of 1941, as he believed that Britain could hold out against the German invasion for the time being and that the United States would not declare war too early. He continued his work with his assistants preparing papers on the topic of operator rings. He also dabbled during this time in astrophysics and published a paper in collaboration with the Pakistan-born Indian physicist and recipient of the 1983 Nobel Prize Subrahmanyan Chandrasekhar entitled *The Statistics of the Gravitational Field Arising from a Random Distribution of Stars*. In his spare time, von Neumann began to revolutionize economics, but this would not come to the forefront until after the war. Of course, von Neumann also coauthored research reports, for example, on the possible errors in successive difference estimation, to which he added three supplements. His openness to collaboration was becoming more and more prominent, something that would serve him well in his future work.

The reports that von Neumann wrote made him the most important expert in the theory of explosions in the United States. A few months after its entry into the war, he also earned a reputation as a master of calculations for complex blasting, for example, collision blasting; Colonel Leslie E. Simon (later General Simon), the commanding officer of the Ballistic Research Laboratory at Aberdeen, particularly admired him. His fame with the Technical Staff caught the attention of the Navy, which quickly recruited him. Von Neumann

Von Neumann with US military officers, with IBM president Thomas J. Watson on the left

preferred naval officers to army officers, because the latter drank only ice water at lunch whereas the former were prone to drink as soon as they landed, and he himself could drink without ever becoming drunk. Later, he was transferred to a mine warfare operations analysis group, with which he worked first for 3 months with the Navy in Washington, not so far from Princeton, and subsequently for 6 months in England, accompanied by his second wife. The two of them were flown across the Atlantic in a bomber.

He was needed in Britain because the Germans had laid magnetic mines in the English Channel in great numbers. Originally these mines exploded and were destroyed immediately in the presence of metal, so that their locations could be easily discovered by a metal trawler, facilitating their safe detonation. As a result, the Germans cleverly installed a mechanism that would prevent them from exploding upon first contact with metal, requiring instead multiple interactions. Each mine was different, and their patterns had not been worked out, so the British had requested assistance from their allies. The problem was an easy one for von Neumann, who completed it easily with the aid of his mathematical skills, preventing the unnecessary sacrifice of innumerable naval officers and soldiers. Among his other work, he derived the equations for conical explosions for the British Navy on the basis of his knowledge of the most destructive oblique shock wave reflections in air and water.

By May of 1943, Washington asked London to send their best experts in explosion theory back. Von Neumann hoped to stay on in England for some time, where he was learning quite a bit about aerodynamics, collaborating with some experimental physics who piqued his interest, and had even developed an interest in computing technology, suggesting that probably he had already met Alan Turing, the British man who established the basic principles for digital computers and who had been one of von Neumann's assistants a few years earlier as a doctoral candidate in mathematics at Princeton University. But sometime in the middle of the year, the United States compelled von Neumann to return. His next task would be an irresistible one, concerning that most important word ever coined by mankind: *nuclear*. But when he first returned, he was still immersed in the topics he had learned in Britain, helping the army to improve its air defense systems, extending the theory of high-altitude explosions and its underwater applications. Shortly everything changed.

Von Neumann was appointed as a consultant to Los Alamos Laboratory in New Mexico—seemingly an inconspicuous title, but which led to results of the highest brilliance. From that time on he held simultaneously four separate posts, at Princeton, with the US Army, with the US Navy, and at Los Alamos, all of them basically full-time positions. He was even wrapping up some work left over from his time in Britain. It was no surprise then that he wrote in a short letter from an unknown address to one of his friends or colleagues that he had been traveling to three or four different places every week since returning from England, now in the southwest at Los Alamos, and that he may even still need to go back to England before Christmas, although he had no way of knowing when he would go nor how long he could stay, but nevertheless that he felt it would be rude not to reply in time. But how much did von Neumann actually contribute to the two atomic bombs that were dropped on Japan?

It was well known that the development of nuclear weaponry has everywhere relied upon collective effort and intelligence, with its first occurrence, in the United States, no exception. The first eureka moment came in 1933 from Zillard, another expatriate from Budapest, who had the idea of a chain reaction: neutrons released in the form of a geometric series could generate amounts of energy beyond the scope of human imagination within a few millionths of a second. At that time, several senior scientific figures, including Einstein, Ernest Rutherford, and Niels Bohr, despised this discovery, and Einstein even remarked in a joke that research into atomic energy was like shooting a small and rare bird in the dark. Zillard, who was working at the London Medical College at the time, was also turned down by the British Admiralty. Meanwhile, the Italian physicist Enrico Fermi had been bombarding every imaginable substance with neutrons in Rome, including metals such as aluminum, iron, copper, and silver and nonmetallic substances such as silicon, carbon, phosphorus, iodine, and even water.

Until Fermi and Guglielmo Marconi, who invented radio, appeared on the scene, Italian science had been in decline since the seventeenth century, when Galileo Galilei had been compelled to submit his guilty plea to the Roman Catholic Inquisition. In the fall of 1934, Fermi bombarded uranium with neutrons, which passed through the nucleus and instigated atomic reactions. Four years later, he was awarded the Nobel Prize for the identification and production of new radioactive elements by neutron bombardment and his discovery of the reactions caused by slow neutrons in this research. Somewhat surprisingly, the government of Benito Mussolini allowed him to travel to Sweden, where he received the prize and promptly boarded a boat set for New York, along with his Jewish wife and their two children. Later that year, three scientists at the Kaiser Wilhelm Institute in Berlin also bombarded uranium with slow neutrons, only to discover that the alleged new element was none other than barium, already familiar as the 56th element. But although the Nobel Prize may seem for this reason to have been misawarded, Fermi also made other important contributions, for example, that a reduction to the speed of neutrons facilitates easy nuclear reactions in irradiated materials. By analogy, it is easier to land a basketball moving slowly in the basket than one moving too quickly and prone to bounce away. His method was quickly adopted by colleagues in various countries.

The three scientists in Berlin were the German chemists Otto Hahn and Fritz Strassman and the Austrian Jewish physicist Lise Meitner, eldest of the three, and another scientific heroine alongside Marie Curie. It was Meitner who coined the word *fission*. If Hitler had not despised nuclear physics on account of his racism and hatred for the Jewish people (he considered it to be some kind of Jewish physics), Germany might well have developed the atomic bomb earlier than the United States. In fact, during that delicate moment, the physicist Heisenberg had solemnly made a report on the possibility of nuclear weaponry to his colleagues in the Army Ordnance Office, but Hitler was unmoved by the prospect. When Hungary was occupied by German, Meitner lost her Austrian citizenship and came to be regarded by the authorities as a German Jew rather than a foreign Jew. Under constant fear of persecution, she eventually escaped to Sweden by way of Denmark with the assistance of Niels Bohr. Her two colleagues who remained behind in Berlin completed their work and secured most of the recognition for it, including the1944 Nobel Prize in Chemistry. Although Meitner, who never married, did not immigrate to the United States, her research results were transmitted there through Bohn, leader of the Copenhagen school of quantum mechanics and Heisenberg's former mentor.

7 John von Neumann, Who Made the World a Better Place

Los Alamos was bursting with talent at this time. In addition to the Hungarian Four, there were also Bohr, Fermi, and Oppenheimer, a native to the United States and the director of the laboratory. There was even a Soviet spy lurking about, the German physicist Klaus Fuchs. What could a mathematician contribute among a team of top physicists? Although von Neumann attached great importance to the intrinsic relationship between mathematics and physics and in fact was himself an outstanding physicist, his main achievements in that arena were entirely theoretical. Specifically, he had contributed to the mathematization of quantum mechanics. But through his efforts, von Neumann became the most respected mathematician among physicists. He guided the design for the optimal structure of the atomic bomb, ensuring that it was small enough to fit within a bomber. He also explored the options for achieving large-scale thermonuclear reactions. According to his view, dynamite like anything else can be replaced by numbers representing the physical elements of an experiment, and in fact these numbers can constitute the entirety of the experiment, provided they are handled correctly, saving not only time but also financial and material resources. It is said that whenever he would enter the laboratory, he would find himself immediately surrounded by colleagues eager to ask his advice on certain calculations.

After a period of collaborative effort, it was determined that uranium-235 and plutonium-239 were the most suitable materials for fission, and it was of these materials that the two atomic bombs dropped on Hiroshima and Nagasaki, respectively, were made (the first atomic bomb produced in China was a uranium bomb, while the recent test explosion in North Korea was of a plutonium bomb). The difference between the two is that plutonium can be separated and obtained through chemical methods, while uranium can be separated only atom by atom, as it has the same chemical properties as its isotopes. For this reason, Japan did not intend to surrender immediately in the aftermath of the bombing of Hiroshima, since scientists reported the second day that it would take a year to produce another identical bomb; the emperor declared unconditional surrender only after the second bomb fell on Nagasaki 3 days later. Plutonium bombs have the advantage that two or three of them can be produced in a month, but at first the bombardment method that was suitable for uranium bombs, which involved shooting one piece of uranium at another, was not suitable for plutonium bombs. Von Neumann played the role of an implosion expert in tackling this problem and personally designed a prism to serve as an implosion device for the plutonium bomb, which proved successful in the first nuclear test (the uranium bomb was not tested).

It is worth mentioning that the United States originally planned to drop atomic bombs on four cities, with the Air Force proposing six candidates: Kyoto, Hiroshima, Yokohama, Tokyo (home to the Imperial Palace), the Kokura Arsenal, and Niigata. Von Neumann was a member of the committee that decided where the bombs should be dropped, and the four cities he recommended were consistent with its final decision to omit Tokyo and Niigata. Fortunately, the Secretary of the Army objected to Kyoto on the grounds that its destruction would incite public outrage as it was a historical capital and a sacred site to both Buddhists and Taoists, potentially altering for the worse the postwar political and economic situation. Yokohama was also excluded because it had already been heavily bombed and because it was too close to Tokyo, and Nagasaki in Kyushu took its place. The Americans realized that the decision to surrender must be made in the capital. On August 6, 1945, the uranium bomb Little Boy was dropped on Hiroshima, and on August 9, the plutonium bomb Fat Man was dropped on Nagasaki, bringing an end to the Second World War. It is not difficult to reason that had the war continued each day would have brought more military and civilian casualties to many countries, including China. The author believes that Hiroshima was among the targets because it was the most remote large city on Honshu island.

Maximum Profits in Business

After witnessing the devastation of the nuclear bomb, Oppenheimer famously found himself quoting a verse from the Indian epic *Bhagavad Vita*: "Now I have become Death, the destroyer of worlds." Einstein too felt deep remorse after the successful bombing of Japan and regarded the letter he had sent to President Roosevelt prompting the creation of the Manhattan Project as the greatest mistake of his life (although the order to drop the atomic bombs was signed by President Truman, after the sudden death of President Roosevelt in April of that year). In 1954, Enrico Fermi, creator of the nuclear reactor, died of cancer. Von Neumann too suffered from nuclear irradiation during the test detonation at Bikini Island. Later, Vyacheslav Malyshev of the Soviet Union and Deng Jiaxian of China suffered similar fates, and although Oppenheimer lived longer than the others, he too made it only to 62. But von Neumann believed that since the theory of the atomic bomb was a discoverable thing, it could not be abandoned to the conscience of a dictator. Many of his relatives had died in fear of the Nazis, some even after they had immigrated to the United States; and many of the scientific advances of mankind, from steam

7 John von Neumann, Who Made the World a Better Place

Von Neumann lecturing at the American Philosophical Society

engines and ships to aircraft, from industrialization to medical experimentation, from guns and ammunition to tank technology, have been the cause of deaths. But these advances have also contributed to the productivity of mankind and the extension of life, whether through saving time and labor or bringing an end to dictatorships, and in this way have also made life more meaningful.

For a person as vigorous as von Neumann, who served three different administrations of the US government, there were too many things he wanted to do related to the livelihoods and constructions of people. Even in 1944, while his colleagues at Los Alamos were at their busiest, he found time to think deeply about economics, and in that year he coauthored a treatise with the economist Oskar Morgenstern entitled *Theory of Games and Economic Behavior*. Upon its publication, it immediately earned the praise of Richard Stone, a British economist from Cambridge University (like John Maynard Keynes) and recipient of the 1984 Nobel Prize in Economics, as the most important textbook in economics since *The General Theory of Employment, Interest, and Money* by Keynes. Game theory was originally a branch of applied mathematics, which later played an important role in both theoretical and applied economics, and found widespread use in politics, military theory, business, law, athletics, biology, and other fields of study. Its influence has been tremendous as a tool for the expansion and refinement of strategic thought; for businessmen, it indicates how to operate in order to maximize profits.

The first person to raise the problem of games was the French mathematician Émile Borel, known to mathematics students for the creation of Borel sets in the theory of functions of a real variable. Borel was also a famous politician and educator, who served at various times as the leader of the École normale supérieure in Paris, a mayor, a member of the Chamber of Deputies, and as Minister of the Navy. In the 1920s, he set down the definitions for game theory, considering optimal games, mixed games, equilibrium games, and infinite games, and also proposed mathematical solutions for solo games and zero-sum two-person games. Two-person games are a fully antagonistic special case of multiparty games, in which by comparison the possibilities of stability and alliance are present. Zero-sum games are set up in contrast with non-zero-sum games; in the former the total payout of each game to the competitors is zero or constant, while in the latter it is variable. In other words, in the first case the gains of one competitor correspond exactly to the losses of another, while in the second competitors can gain or lose simultaneously.

In 1928, von Neumann, at that time still an unpaid lecturer at the University of Berlin, published the first significant paper in game theory, *On the Theory of Parlor Games*, in which he used a matrix representing bargaining power to analyze a zero-sum two-person game. The Minimax theorem, which appears in this paper, later became the cornerstone and central theorem of game theory. As an application, von Neumann discussed the problem of collaborative games, with special consideration given to a zero-sum three-person game in which two of the parties band together. To this end, he introduced the concept of characteristic functions in mathematics, clearly set out a game plan for multiple competitors, and proved under certain additional conditions that a solution to the multiparty game problem exists and is unique. As an example, von Neumann's theory indicates that the success of say the economic behavior of Ford company is not determined entirely by the market but also takes into account the strategies implemented by General Motors, Japan, Germany, and other automakers.

In 1932, von Neumann delivered an unscripted report at a mathematics seminar at Princeton on the topic of several equations in economics and a generalization of Brouwer's fixed point theorem. In his report, von Neumann pointed out the solutions to economic problems from a mathematical perspective, which described an expansive new economic model: all goods produced at the lowest possible cost and in the largest possible quantity, an idealized model that automatically achieves dynamic equilibrium once the maximum growth rate is achieved. Four years later, von Neumann intended to deliver these remarks a second time at a mathematical conference in Vienna but had to change his travel plans due to the breakdown of his marriage. Since he was unable to travel, he hurriedly scrawled nine pages in German in a Paris hotel, and they were included in the conference proceedings, which were subsequently published but did not receive any special attention. In 1945, this article was translated into English and republished in the United Kingdom under the title *Model of General Economic Equilibrium*. Around half a century later, this paper gained recognition as the most important paper in mathematical economics.

Von Neumann introduced to economics the theory of linear and nonlinear programming, and the science of modeling for future development, enabling people to better understand the activity and inactivity of planned and market economies. To date, at least six recipients of the Nobel Prize in Economics have acknowledged the influence of von Neumann on their work. These are Paul Samuelson, Kenneth Arrow, Leonid Kantorovich, Tjalling Koopmans, Gérard Debreu, and Robert Solow. A further five laureates received the Nobel Prize for work and directly developed or applied the theory of games introduced by von Neumann: John Harsanyi, John Nash, and Reinhard Selten who shared it in 1994 and Robert Aumann and Thomas Schelling, who shared it in 2005. These various economists hail from the United States, Britain, Germany, the Soviet Union, and Israel. Even in Japan, there are economists who have admired von Neumann and followed the model he advocated: for dynamic equilibrium to exist, it is necessary to maximize production. After the war, the national policy of Japan was to strive to double real income within 10 years. At that time, some Western economists concluded that this would lead to serious inflation, but they turned out to have been unnecessarily cautious, and the economy of Japan developed along the track of a virtuous cycle.

In 1938, the German economist Oskar Morgenstern came to teach at Princeton University, giving von Neumann an opportunity and room to expand upon his theory and also leading him to become interested in economic behaviors such as the exchange of goods, market control, and free competition. After several years of collaboration, they completed their economic masterpiece of more than 600 pages. Nevertheless, many postwar economists disagreed with his theory and even harbored some resentments toward him, in part due to various misunderstandings and in part because he was an outsider to the discipline of economics. But the passage of time and practice has gradually eliminated these misunderstandings, and today von Neumann is universally recognized as the founder of game theory and a pioneer of mathematical economics, which is an important branch of modern economics. Samuelson remarked that although he only scratched the surface of economics, von Neumann was unparalleled, and the field was no longer what it used to be before him.

Who Made the World a Better Place

Ever since Newton and Leibniz had invented the calculus and brought about the mathematization of physics, scientists were in greater and greater need of computation tables. In addition to the familiar tables of logarithms and trigonometric functions, there appeared various special tables temporarily developed in the ordinary course of research. Leibniz lamented time spent like a slave on the drudgery of calculations and for this reason invented a calculating machine somewhat like a mechanical abacus which could carry out addition and multiplication with a shake of its wheels. This was an improvement upon an earlier mechanical adder invented by Blaise Pascal, but not enough to satisfy Charles Babbage, a whimsical British mathematician of the nineteenth century, who tried unsuccessfully to improve upon these designs using the most fashionable steam technology of the time. Nevertheless, he realized the mechanical computers should be based on precise mathematical logic, and not long afterward the self-taught Irish mathematician George Boole invented a new form of mathematics to meet this challenge, known today as Boolean algebra.

By the middle of the twentieth century, the situation had changed yet again. At Los Alamos, the large number of computational tasks generated during the pursuit of nuclear fission prompted von Neumann to pay attention to the development of electronic computers. In the same year that *Theory of Games and Economic Behavior* was published, he met by chance with Herman Goldstine, who had participated in the design of the first electronic computer *ENIAC*, on a railway platform in Aberdeen. Von Neumann was on his way to Los Alamos at the time and took immediate note of the remarks of his colleague, and over the next 2 years he participated personally in the design of two computers at the University of Pennsylvania and the Institute for Advanced Study: *EDVAC* and *IAS*. He observed that the main flaw of *ENIAC* was that it adopted the plug-in type of previous electromechanical computers and established the most important structural principle within the computer, the stored-program principle, which comprises five parts: the processing unit, control unit, memory, external storage, and input and output devices.

These two devices had many advantages over *ENIAC*, the most important of which consisted of replacing decimal with binary, the representation of programs and data by binary digits (binary notation had been invented centuries earlier by Leibniz but played no role in his mechanical calculator), and the change in the programming mechanism from external plug-ins to memory, which meant that there was no need to change the circuit board for each

7 John von Neumann, Who Made the World a Better Place

Von Neumann in front of the electronic computer he helped to develop

new calculation, but only the program. Electronic computers constructed according to the stored-program principle are called von Neumann machines or von Neumann architecture, and this structure is still in use today, for which reason von Neumann is referred to as the father of the electronic computer. Alan Turing, another genius in the field of computing, made major contributions to ideal computers and artificial intelligence and is widely known for his untimely death in connection with persecution for his sexual orientation and the Turing Award in computer science. Von Neumann devoted more of his energy to *IAS* since he served himself as the director of computer science at the Institute for Advanced Study. Their machine succeeded at last in 1951, running hundreds of times faster than *ENIAC*.

Although the name von Neumann is associated with computer design, he himself was not so interested in the design and manufacturing of computers, but rather in the use this new tool could offer to create a new world for scientific computing. In 1950, von Neumann led a team of researchers in weather forecasting to use *ENIAC* to complete the first successful calculation in the history of numerical weather prediction. With the increasing computing needs in weather forecasting, and other scientific and engineering fields, computational methods were as essential as computer hardware for the improvement of computing speed. This brought into being a new branch of mathematics in addition to pure mathematics and applied mathematics: computational mathematics, which not only designs and improves upon various methods of numerical calculation but also studies such related issues as error analysis, convergence, and stability. Von Neumann is without a doubt an early founder of this discipline.

Throughout history, mathematicians from around the world have contributed convenient methods for numerical calculation. The classical methods, however, are not necessarily identical with the optimal methods for computers, and some methods which appear to the human eye bewilderingly complex are actually easy to implement as a computer program. With respect to the methods and techniques for computer calculations, von Neumann made many important contributions. He created in succession more than ten computational protocols, including the calculation of matrix eigenvalues, matrix inversion, the evaluation of multivariate functions, and random number generation, widely used in the industrial sector and government planning. Particularly noteworthy was his collaboration with the Polish-American mathematician Stanisław Ulam to invent the Monte Carlo method, a new form of calculation in which approximate solutions to problems are obtained through artificial sampling, converting mathematical questions into probability models subject to random simulation by computer programs. This method enables, for example, pollsters to produce more accurate predictions about upcoming presidential election results on the basis of only a small amount of sampling or random sampling.

The Monte Carlo method embodies the ability of the electronic computer to process large quantities of random data and represents a pioneering moment in the development of new algorithms for the computer age. There are two steps for the solution of practical problems: the simulation of random variables generating various probability distributions and the use of statistical methods to estimate the digital characteristics of the model to obtain numerical solutions. Von Neumann named it after a city of casinos, adding to its mystique. It has also found widespread application in the field of financial engineering, including the pricing and transaction risk estimation of financial derivatives, options, futures, swaps, and so on. The number of variables, or dimension, involved is as high as the hundreds, even thousands, which has been referred to as the curse of dimensionality. But the Monte Carlo method has the advantage that its computational complexity does not depend on the number of dimensions. In the 1970s, the Chinese mathematicians Hua Luogeng and Wang Yuan replaced the random number sequences involved in the method with a deterministic hyperuniform sequence, proposing the so-called quasi-Monte Carlo method. In some applications, the Hua-Wang method is hundreds of times faster than the ordinary Monte Carlo method, with computable accuracy.

If von Neumann were alive today, he would feel deeply gratified to see the surge in the number of computers in the world, and the improvements to their capabilities, which have proved indispensable for companies, governments, schools, and families, in the skies and on earth. His daughter, the economist Marina von Neumann Whitman, remarked that he would have surely been surprised to learn that the company for which she worked, General Motors, produced and used millions of computers every year, including some eight million distributed across every car produced by the company, and he might have chuckled at parents who rail against children spoiled by video games, since he had a childlike and playful personality. But we may speculate that while von Neumann would surely have been amazed at the benefits brought to society and to humanity by computers, he might also feel frustrated that they have failed to achieve even greater achievements in science.

If His Life Had Been Longer

In 1992, the British economist Norman MacCrae, who wrote a biography of von Neumann, wondered if it would have made a big difference in our lives if he had lived a life as long as that of an average scientist, which would have seen him still alive at that time. His own answer to this question was affirmative. Judging from unpublished notebooks from his later years, von Neumann had his own vision for the future of science. In fact, at the end of his life, he was still exploring some questions that other scientists had never considered, for example, what could be learned from the human nervous system for application to electronic computers? Such lines of thought recall the questions he asked at those lunchtime family gatherings in his childhood. On his deathbed, von Neumann worked on a book entitled *The Computer and the Brain*, published posthumously, according to which the computers and robots of the future should respond more efficiently to their environments, and the next generation of computers should be capable of self-reproduction subject to laws of evolution and the survival of the fittest.

In the first half of the twentieth century, von Neumann personally participated in three revolutionary breakthroughs—the scientific understanding of the atom, the mathematization of quantum mechanics, and the subsequent development of electronic computers—and made outstanding contributions to each. The lectures and notebooks he left behind show that he hoped to play the same role in three possible major breakthroughs for the future—the scientific understanding of the brain, the understanding of cells and genetics, and

Tomb of von Neumann (photograph by the author, Princeton)

the understanding and management of the natural environment—by which he meant, for example, to control the weather rather to predict it, to convert Iceland into Hawaii. He hoped also to make precise the vague discipline of economics. According to his vision, all concepts of numbers in the computer age should also be reestablished. Unfortunately, of these hopes, only genetics has seen satisfactory progress so far, and this was based on a discovery made during his lifetime, specifically the double helix molecular structure of deoxyribonucleic acid or DNA. As von Neumann predicted, genes consist of information stored in a way similar to computer storage.

If von Neumann had lived longer, he would have been as excited by molecular biology as he was by quantum mechanics, and he would have looked forward to mathematizing it. Interestingly, his only grandchild is a molecular biologist at Harvard Medical School. As for the development of other scientific fields, it has not been as successful or as fast as von Neumann expected, perhaps because the world lost him prematurely. In the third-century BCE, the Greek mathematician Archimedes used a giant ballistae to fire stone projectiles weighing 250 kg each, destroying a Roman fleet. Like Archimedes, von Neumann used his mathematical skills to help the United States win the final victory of the Second World War. In 1956, von Neumann received the first Albert Einstein Commemorative Award and the Enrico Fermi Award. The latter award is given to those who have made outstanding contributions to the scientific understanding of the atom; its namesake was the first recipient, and von Neumann the second.

In 1957, as distant China was launching its large-scale anti-rightist movement, von Neumann's life was coming to its end. Bone cancer cells caused by nuclear radiation had spread in his body (he died 9 years earlier than Deng Jiaxian). Von Neumann felt early on that the most intelligent people he encountered were often either Jewish or Chinese. In his later years, he praised the Chinese language as the language of poetry in his notebooks. In 1937, von Neumann learned about the current situation of Chinese mathematics from the American mathematician and inventor of cybernetics Norbert Wiener and developed a desire to visit China. Wiener had lectured for a year at Tsinghua University and wrote to its principal Mei Yiqi and the head of its Department of Mathematics Xiong Qinglai. Unfortunately, the Marco Polo Bridge Incident occurred 2 months later, the Japanese invasion broke out, and this wish came to nothing. When Wiener and the French mathematician Jacques Hadamard had visited China, it had caused a stir in the mathematical community there, and no doubt if the versatile John von Neumann had come to China, the impetus of his visit would have been inestimable, and he himself may have gained tremendous inspiration from a novel journey to the East.

Von Neumann died on February 8 at the Walter Reed Army Hospital in Washington at the age of 53. As he lay dying, the US Secretary of Defense and Deputy Secretary of Defense; the commanders-in-chief of the Army, the Navy, and the Air Force; and other military and political officials gathered at his bedside to hear his last suggestions and extraordinary insights. General Lewis Strauss, then chairman of the Atomic Energy Commission, witnessed the scene and referred to it years later as the most dramatic sight he had ever seen and the most moving tribute to a wise man. President Eisenhower had previously personally awarded a special Medal of Freedom to von Neumann, who was in a wheelchair by that time. At the same time, FBI agents in disguise monitored the ward day and night, worried that the comatose genius might reveal national military secrets. General Strauss could not have imagined at that time that half a century later the hospital would come to serve primarily soldiers wounded in the wars in Afghanistan and Iraq. At dusk, the afterglow of the setting sun fell on both sides of the Potomac River and passed through the windows of the Army Hospital, the last moment of glory before the sunset, when one of the greatest and most active minds of the twentieth century stopped thinking.

Spring 2010, Caiyunju, Hangzhou

8

Paul Erdős: A Narrowly Missed Opportunity

A mathematician is a machine for turning coffee into theorems.
—*Alfréd Rényi*

A Trapeze in the Air

The protagonist of this chapter should be the envy of any lover of travel. Apart from salesmen, tour guides, diplomats, and flight attendants, he may have spent more time in the air than anyone. He had no fixed occupation nor income but lived in hotels and ate in restaurants every day, his bills all paid for by others. He was a pure child prodigy with an unparalleled mind who later became the most prolific mathematician in history and made groundbreaking contributions to many diverse fields of mathematics. This was the Hungarian mathematician Paul Erdős, the subject of a biography entitled *My Brain Is Open* by Bruce Schechter, which I have read carefully in its Chinese translation for Shanghai Translation Press.

Alfréd Rényi (1921–1970) was a Hungarian mathematician who specialized in number theory, among other fields, after whom the Institute of Mathematics of the Hungarian Academy of Sciences is named. Rényi and Erdős coauthored 32 papers, putting him in 4th place among authors with Erdős number equal to 1.

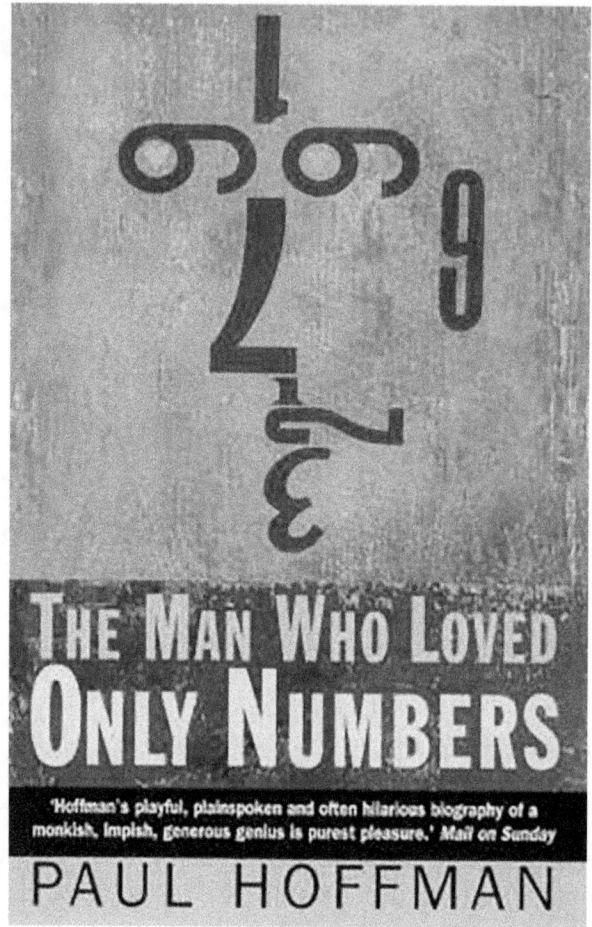

Biography of Erdős, *The Man Who Loved Only Numbers*

In the early 1960s, relations were close between China and Eastern Europe, and Erdős was invited to Beijing, where he met the mathematician Hua Luogeng. Wang Yuan, who participated in the translation of this biography, was Hua Luogeng's assistant and collaborator at the time but had gone to Shanghai to visit friends. This detail does not appear anywhere in the book or its endnotes but was told to me personally. Nearly 30 years later, Erdős came to China a second time, and this time not only did he meet Wang Yuan, who was already an academic at the Chinese Academy of Sciences but also traveled to Jinan to attend a conference. I was a graduate student in Jinan at that time, and one gloomy autumn afternoon, Erdős and I spent some time discussing mathematical problems behind the closed doors of a suite at the International Student Building of Shandong University.

I remember that what Erdős proposed to me at that time was the mean-value estimation of certain types of number theoretic functions, which I did not complete, but rather I developed the mean-value estimation for another type of number theoretic function. This topic was brought to me by Professor Pan Chengbiao of Peking University, the brother of my mentor Pan Chengdong. Erdős was the pioneer of this type of estimation, and I managed to improve on his results, or more precisely the results in collaboration with the Indian mathematician Krishnaswami Alladi. This work was not only published early in *Chinese Science Bulletin* but also earned me my master's thesis in advance, with first place in the graduate thesis award of Shandong University. Nevertheless, I missed out an earning an Erdős number of 1 (which means having published a paper in collaboration with him), one of my lifelong regrets.

In the first spring of the new century, I flew from Australia to Kyushu Island in Japan to attend the Second China-Japan Number Theory Conference. Erdős had passed away by this time, but Alladi, then a professor at the University of Florida, was in attendance; in addition to the academic reports of the conference, he was invited to deliver a speech to local college students, in which he told the story of his compatriot Srinivasa Ramanujan and the story of Paul Erdős.

Turning to the third page of *My Brain Is Open*, I saw again the name Alladi, and it turns out that he was one of the many young mathematicians who benefited from Erdős's help. In 1974, when Alladi was still a student at the University of Madras (where Ramanujan had been a researcher), he carried out research on some problems in number theory and came up with some profound insights that even his father, who was the director of the Institute of Mathematical Sciences, Chennai (Matscience), could not resolve. At the suggestion of a friend, he wrote to Erdős; since Erdős traveled for years at a time, he sent it to the Hungarian Academy of Sciences.

To his surprise, Erdős replied quickly, saying that he would soon be visiting Kolkata to present some lectures and asking whether or not Alladi would like to meet there. Unfortunately, Alladi had to sit for an important examination, so he sent his father to introduce Erdős to the work of his son; Erdős sincerely remarked that he was not interested in the father, but very interested in the son.

Erdős decided to meet with the young man and rearrange his itinerary, in which his next stop was in Australia, to accommodate a short stay in Madras (now Chennai), more than 1300 km from Kolkata. Alladi was a bit nervous to meet his ideal mathematical mentor at the airport, but Erdős recited a ballad about Madras to him, putting him greatly at ease, and they moved on to a discussion of number theory. Erdős was impressed by Alladi's talent and immediately wrote a letter of recommendation for him.

In less than a month, Alladi obtained a Presidential Scholarship to attend the University of California, Los Angeles, paving the way for this future mathematician. Erdős donated the entirety of the remuneration he received for his lectures in Madras to the widow of Ramanujan; although Erdős had never met this Indian mathematician of genius, he had been moved as a student by Ramanujan's beautiful equations, which led to his interest in India and his lifelong support for Indian mathematicians.

Childhood in Judapest

Paul Erdős was born in Budapest on the Danube River on March 26, 1913. Like Harold Bloom in *Ulysses* by the Irish author James Joyce, his parents were Hungarian Jews; although the foreign policy of Israel for a long time has not garnered unanimous acceptance from people around the world, the outstanding contributions of the Jewish people in economics, science, culture, and the arts are obvious to all, and in the Hungarian scientific community alone, the twentieth century produced such Jewish luminaries as John von Neumann, the inventor of digital computers and game theory; Edward Teller, father of the hydrogen bomb; Theodor von Kármán, father of supersonic flight; and George de Hevesy, who invented isotopic tracing technology.

Nor was Hungary lacking in artistic talent: there were the pianists Georg Solti and George Szell, the conductors Antal Doráti and Eugène Ormandy, the composers Béla Bartók and Zoltán Kodály, the master of design László Moholy-Nagy, the entertainment tycoon William Fox, the film producer Adolph Zukor, and so on. The preeminence of the Jewish community in the cultural life of Budapest was such that in the twentieth century it was sometimes affectionately referred to as Judapest.

Erdős's parents had been classmates in the mathematics department at the University Pázmány Péter, and his father taught at a middle school after their marriage. At that time, after a half a century under the Austro-Hungarian dual monarchy, Hungary had reached the peak of its economic and cultural glory. But just as his mother was admitted to the hospital and preparing to give birth, a terrible scarlet fever was sweeping through Budapest. By the time she returned home from the hospital with Paul, his two elder sisters had died, and the heartbroken parents poured their love and energy into this gray-eyed boy.

When Erdős was just 3 months old, Archduke Franz Ferdinand, heir presumptive to the throne of Austria-Hungary, was assassinated in Sarajevo, triggering the start of the First World War; Austria-Hungary declared war on

8 Paul Erdős: A Narrowly Missed Opportunity

Outdoor cafe in Budapest (photograph by the author)

Serbia, and Russia quickly became involved, declaring war in turn on Austria-Hungary. This was the end of the golden age of Hungary. Erdős's father was also drafted into the army. He was soon captured by the Russian army and spent 6 years behind bars in Siberia.

His situation reminds me of the Hungarian poet Sándor Petőfi, who was also captured by the Russian army in the mid-nineteenth century and sent to Siberia, where he died of tuberculosis 7 years later. Fortunately, Erdős's father survived the Siberian labor camp and returned to Budapest to find that Paul was already a beautiful little boy who had embarked upon his family education in the Jewish tradition. Mathematics naturally was among the core subjects, but foreign languages played an equally important role, and in addition to German, Erdős learned French and English from his father, who had studied them in Siberia to ward off the severe cold and hunger.

Like almost all Hungarians, however, Erdős spoke English with a strong accent, which I can still remember clearly. Apparently almost every documentary about him includes subtitles for his speech. As a middle school teacher, the mathematics his father could teach Erdős naturally consisted of number theory concerning the properties of integers and especially the prime numbers that serve as their atomic elements. And like most mathematical prodigies, Erdős developed an irresistible interest in prime numbers, spanning such topics as the fact that there are infinitely many of them, which appears in Euclid's *Elements*, to the gaps between consecutive primes, which includes the famous twin prime conjecture.

As is the case with many child prodigies, Erdős was not equipped with very robust basic survival skills; at the age of 11, he finally learned to tie his own shoes, entered school for the first time, and suddenly found himself in the sixth grade all at once. Although the strict classroom discipline of the formal schoolroom inhibited his independent mind, nevertheless Erdős obtained among the best grades of his class, earning an A in every subject but painting. His favorite class at the time was history, a hobby he retained throughout his life. He was prompted to turn his interest to mathematics by a magazine entitled *Middle School Mathematics*, which provided some challenge questions and included photographs of the winners.

Many issues of this magazine were dedicated to the field of number theory, and the effect of his early education with his father shone through. Soon his own photograph appeared, and this magazine remained his companion through middle school. Despite the widespread antisemitism of the times and quota laws limiting Jewish enrollment in universities to 6% of total enrollments, Erdős secured a place at the University of Budapest, where he encountered many faces that he had only known vaguely via this magazine. His mathematical ship began to take sail.

First Trip to England

In September of 1934, Erdős boarded a train and left Hungary for the first time, the first of his countless mathematical journeys. A few months earlier, he had received his doctorate in mathematics from University Pázmány Péter, the alma mater of both his parents, and the University of Manchester in the United Kingdom had offered him a scholarship of 100 pounds. He was not able to take pleasure in the journey, however, but rather he felt tired and was not even able to easily manage arranging three meals daily and other trivial matters on the train. He took pleasure only in the exchange of mathematical techniques; passing through Switzerland, he had the first opportunity to open his brain and pay a visit to a mathematician in Zurich.

The morning of October 1, Erdős traveled by train and to Cambridge, a day he never forgot. He spent no time visiting this world-famous university town but proceeded immediately to Trinity College to open his brain again to an academic discussion with two mathematical colleagues who had come to welcome him. Later, when they had lunch together, his colleagues discovered that Erdős did not even know how to put butter on sliced bread.

A young Erdős

After this brief visit to the University of Cambridge, Erdős continued by train north and then west to Manchester. This city is now famous for football, although at that time it had only won two First Division championships and one FA Cup, both in the early twentieth century. The research center for mathematics at Manchester University on the other hand had long been famous, and at that time it attracted many foreign visitors to give lectures or to collaborate on research due to rising tensions on the European continent.

At that time, intellectuals from the continent were not yet inclined to immigrate across the Atlantic to the United States. Professor Louis Mordell, the chair of the Mathematics Department at the University of Manchester, was himself an American, who had saved money to study in England after graduating from high school and eventually established himself after quite some hard work as a well-known number theorist. The solution of a conjecture named after him later earned a Fields Medal for Gerd Faltings, and in fact, prior to the proof of Fermat's last theorem, this result, now known as Faltings's theorem, was widely regarded as the most important achievement in number theory in the twentieth century.

During his time in Manchester, Erdős cowrote a paper on combinatorics with Richard Rado, a German-born mathematician, and Ke Zhao, one of

Mordell's Chinese students, which contained the theorem that later became famous as the Erdős-Ko-Rado theorem (Ke Zhao is also known in English as Chao Ko). There was not much interest in combinatorics in the mathematical community at that time, however, so this work was not published until 1961, when it immediately became a classic. Professor Ke Zhao was my fellow native of Zhejiang province. At his 80th birthday in Chengdu, Sichuan, I had the opportunity as his junior to enjoy a conversation with him in authentic Taizhou dialect.

After Ke Zhao obtained his doctorate in Manchester, he returned to China and served as a professor at Sichuan University and Chongqing University. It was at his invitation that Erdős first came to China. As the only two number theorists, Ke Zhao and Hua Luogeng were elected as the first academic members of the Chinese Academy of Sciences. Unfortunately, Ke did not achieve the same heights in his research after returning to China due to environmental reasons and other factors, but he nevertheless cultivated a group of great mathematical talents and opened the way for a new world of mathematics in southwestern China.

During his 4 years in England, Erdős was never content to stay in one city, and in fact he rarely slept in the same bed for a full week, traveling constantly between Manchester, Cambridge, Bristol, London, and other university towns with his brain open. During this period, the work of the youthful Erdős revealed his unique personality, playful, sensitive, and original. As one example, he conjectured that a square can be divided in a certain way into several squares of different sizes. It was not until more than 40 years later that it was proved that the minimum number of such squares is 21.

During the Second World War, a young British man named Bill Tutte was recommended to participate in the secret military plan led by Alan Turing on the basis of his work on this conjecture by Erdős. He turned out to be the right person for the job and successfully deciphered the code used by German submarine captains, enabling the Allied forces to intercept and destroy the military supply ships of the enemy, greatly reducing the length of the war. This was probably the best recompense to Britain for inviting Erdős to visit.

Princeton, On the Shore

October 24, 1929, was Black Monday, when stocks plummeted in New York, leading to a decade of global economic panic, which did not let up until the outbreak of the Second World War, which stimulated the recovery. Just over a

8 Paul Erdős: A Narrowly Missed Opportunity

Erdős delivering a report

month earlier, the head Bamberger's, at that time the fourth largest retail chain in the United States, had used his keen insight to transfer the company. Subsequently, perhaps out of guilt, Louis Bamberger and his sister paid a visit to the famous educator Abraham Flexner, who advised them to abandon their intention to establish a medical school.

Flexner envisioned an intellectual Eden after the spirit of Plato's Academy, a safe haven where scientists and scholars could make the world and its phenomena the subject of their experiments without forcible involvement in the whirlpool of current events. This whirlpool of course referred to the disasters caused by Nazi Germany and fascism, which were spreading out of control around the world.

This was how the Institute for Advanced Study at Princeton came to be, with Einstein invited as its leading professor. Every tenured professor there is exempt from the ordinary worries of human beings, including such household mundanities as paying utility bills, the filling out of various forms such as grant applications, even publishing papers and reporting on their work to their supervisors, and so on. In brief, anyone who enters this institution enjoys its complete trust and can pursue any line of research according to his or her interests.

In fact, mathematicians and theoretical physicists spent much of their time strolling along the neatly manicured lawns, chatting over coffee, and playing endless games of chess in the common room. Nevertheless, they made extraordinary contributions, often the best work of their lives. For example, the British mathematician Andrew Wiles published no papers for 7 years until he finally completed his proof of Fermat's last theorem. These phenomena enshrine Flexner as a contributor to human civilization no less important than another doctor of the same generation, the Austrian psychoanalyst Sigmund Freud.

In the summer of 1938, Erdős returned to Hungary from England for his summer vacation. In early September, Hitler, who had just annexed Austria, demanded further the annexation of the Sudetenland, a German-speaking region of what was at that time Czechoslovakia. Erdős was shocked, but at this moment Princeton extended to him an olive branch, and invited him to come there as a guest lecturer. At the age of 24, Erdős bid a hundred farewells to friends and relatives, many of whom later died in the war, and boarded a train with a southward detour through the Pannonian Basin, Italy, and Switzerland, before reaching Paris and continuing on to London, where he set out on the *Queen Mary* for New York and eventually New Jersey, a solid step toward a tour of the world.

In his later years, Erdős believed that his first year after arriving at Princeton was the most successful in his academic career. He proved, for example, that the product of any number of consecutive integers cannot be a perfect square, which reinforced again people's belief in the orderliness of the structure of numbers. As another example, Erdős and the Polish American mathematician Mark Kac obtained the Erdős-Kac theorem, which states that the number of different prime factors of an integer less than some given integer N follows the same distribution as the number of appearances of heads in N coin tosses, which by contrast introduces the mathematics of chaos into the rules of ordinary integers.

However, the sort of problem that attracted Erdős and to which he was suited was not taken seriously at the time, since they were far removed from recent trends in the development of mathematics. His view was that there still exist countless treasures to be found in the ordinary language of mathematics, why not exploit it and seek them out? Besides, these problems form the most beautiful part of mathematics. One of his collaborators remarked that Erdős's imagination and skill were so profound that he could create a never drying stream without venturing too far, while others, because their imagination was not as deep and their skills not refined enough, need to go through more mathematics to arrive at new ideas and new theorems. In any case, Princeton did not renew Erdős's appointment, which he resented.

The door to the Garden of Eden closed behind Erdős, and he was obliged to embark upon a new mathematical journey; from this point on he became a true wanderer. But Erdős was open-minded and continued to visit Princeton frequently after the war. It was at Princeton that he provided an elementary proof for the familiar prime number theorem. It was mentioned that Atle Selberg, a Norwegian mathematician, independently provided an elementary proof of this theorem in the same year, and it was this result that especially earned him a Fields Medal. Erdős received some comfort 34 years later, when we has awarded the Wolf Prize in Mathematics for lifetime achievement, alongside Shiing-Shen Chern. Selberg passed away in the summer of 2007. One year earlier, in an interview with two Norwegian mathematicians, he gave a detailed discussion and comparison of the work that he and Erdős had separately completed at that time.

The Don Juan of Mathematics

One of Einstein's assistants, Ernst Straus, was obsessed with mathematics and once gave voice to his concern that a person may exhaust all of his energy on a certain problem without ever discovering the key to it. Einstein likewise believed that he himself did not become a mathematician because the field was too full of beautiful and difficult problems. Erdős on the other hand never looked back but dove deep toward the temptation that frightened Einstein, without ever falling into the quagmire of irrelevant matters. These two remind me, respectively, of two geniuses of the seventeenth century, Pierre de Fermat and Isaac Newton. The former devoted himself wholeheartedly to pure mathematics and problems in number theory, while the latter invented calculus, the three laws of motion, and the law of universal gravitation, becoming the most important scientist in history.

Straus believed nevertheless that in the search for truth, there is a place for both a Don Juan such as Erdős and a Galahad such as Einstein (the former being a fictional lothario, the latter a legendary knight). Unfortunately, after I had worked out the mean estimation problem discussed above, a senior colleague, in the spirit of the traditional mindset, told me that this kind of work associated with Erdős amounted to a small problem. This friendly advice prevented me from going in the direction I knew myself to be best suited to. It was not until the proof of Fermat's last theorem appeared that many knowledgeable people, including Professor Wang Yuan, realized that number theorists would be well served by returning to the original mathematical problems with which Erdős was so enamored.

Erdős was an ascetic who abandoned worldly pleasures and material pursuits to live a life of exhaustive effort that was not easily understood by other people. Like Pascal and Newton, he never married nor even fell in love throughout his life; this was not a by-product of mathematics however, but due to congenital psychological reasons: a total lack of interest in sex, indeed an aversion to even the slightest physical contact. When a stranger would venture to shake his hand, Erdős could at most manage to brush them slightly with his soft hand, and even this made him uncomfortable to the point that he felt compelled to wash his hands again and again throughout the day. It was not the case that no woman took an interest in him, but at the critical moment he would run away.

What was it about mathematics that so intoxicated Erdős and led him down this haggard path? In addition to his predilection for playfulness, agility, and originality already discussed, the omnipresent challenges of mathematics were a stimulant to his nerves not unlike opium. His brain was always open, his ears always alert. Two of his works, the elementary proof of the prime number theorem and the generalization of the Seven Bridges of Königsberg problem, came to him by hearsay and on the phone.

Bertrand Russell was a mathematician and philosopher who by contrast married four times and engaged in many affairs throughout his life, including secretly to the first wife of poet T.S. Eliot. He received the Nobel Prize in Literature partly because his writing style was elegant and appealed equally to refined and popular tastes. In his youth, Russell was also fascinated by the world of numbers and even wrote a poem of praise to Pythagoras and his certainty in numbers.

Russell was born into the aristocracy, and his grandfather had served twice as the British Prime Minister. Both of his parents died when he was 3 years old, and Russell grew up under the strict discipline and puritanical training of his grandmother; he even considered suicide in his youth. It was through mathematics that Russell acquired fame and escaped the loneliness and despair of his adolescence. Although later he turned his attention to philosophical research, Russell benefited from his familiarity with mathematics throughout his life. His philosophical masterpiece, which he coauthored with Alfred North Whitehead, was the *Principia Mathematica*, which carried forward the program and ideas of logical positivism, raising new goals and new questions for philosophical research.

Georgia Tech 1996 III 19

Georgia Institute of Technology
School of Mathematics
Atlanta, Georgia 30332-0160
USA
404•894•2700
FAX 404•853•9112

Dear Professor Melfi,

I just read your interesting paper on practical numbers. Denote by $f(x)$ the number of integers $n < x$ for which the integers n, $n+2$, $n+4$ will all be practical. I think

$$(1) \qquad f(x) > \frac{x}{(\log x)^c}$$

will be true for large c and $x > x_0(c)$. (You only got $f(x) > (\log x)^c$)

Also the following result should hold: for every t there will be t consecutive integers $m+1, m+2, \ldots m+t$ which all are "nearly" practical i.e. there is a constant $l(t)$ depending only on t so that for every i, $1 \le i \le t$ every integer u, $l(t) < u \le m+i$ is the sum of distinct divisors of $m+i$, also the number of these "nearly" practical numbers $n < x$ will be greater than $\frac{x}{(\log x)^{\alpha}}$ where c depends only on $l(t)$ (or rather on t)

I do not see at the moment how to prove these conjectures.

At present I am at the University of Memphis Department of Mathematics Memphis Tennessee 38152 (c/o Prof Ralph Faudree) or you can write to Prof Jean Louis Nicolas Univ of Claude-Bernard Lyon Villeurbanne France or to my Budapest address

Kind regards
Paul Erdős

An Equal Education and Employment Opportunity Institution — A Unit of the University System of Georgia

A handwritten letter by Erdős

The Hungarian mathematician John von Neumann, a contemporary of Erdős, was another dynamic figure, the archetype of a generalist, who made outstanding achievements in mathematical logic, set theory, the theory of continuous groups, ergodicity theory, quantum mechanics, and operator theory. He was also the father of modern electronic computing and game theory, with a huge influence on the field of economics. Even Erdős could not help but admit that von Neumann's speed of comprehension and reaction were astounding. In addition to his quickness of thought, von Neumann was stylish and charming and enjoyed sports, cars, physics, and women, with a fondness for limericks and pornographic stories. He was by no means averse to noise, food, wine, and money.

I have mentioned here the examples of Russell and von Neumann simply to illustrate that the personalities of mathematicians vary from person to person, and are not necessarily related to the characteristics of mathematics itself. On the other hand, it may be that a mathematician such as von Neumann derived his ideas from experience, and his life was generally like this; this was not the case for Erdős. At least from my perspective, his mathematics sprang directly from a brain that was always open. In the autumn of 1996, Erdős suddenly died of a heart attack while lecturing in Warsaw; that magical brain ceased its working, and a great flow of mathematics came to a halt. It may be a thousand years before humans encounter another such brain.[1]

February 2004, Hangzhou (first draft)
Completed August 2008, Cambridge

---

[1] Erdős had the idea of *The Book of Mathematics*; whenever he found a beautiful proof, he would praise it has having come from this heavenly book. This was the model for the book *Proofs from the Book*, which has not only since been published but also exists in Chinese translation in several versions, under the title 天书中的数学证明, meaning *Mathematical Proofs in the Heavenly Book*.

9

Mathematicians and Poets

Mathematics is the art of giving the same name to different things. Poetry is the art of giving different names to the same thing.
—*Henri Poincaré*

Mathematicians and poets exist in our world as uncanny prophets. The difference between them is that poets are thought to be arrogant because they tend to be proud and lonely by nature, while mathematicians are thought to be unapproachable because they exist on a transcendent plane. Thus, in art and

Novelist William Faulkner

Portrait of the poet Stéphane Mallarmé (by Édouard Manet)

Self-portrait by Edgar Degas

literary circles, poets are often considered to be socially inferior to novelists in the same way that mathematicians are considered socially inferior to physicists in scientific and technological associations. But these things are only superficial.

9 Mathematicians and Poets

"I'm a failed poet," the novelist William Faulkner said humbly in his later years. "Maybe every novelist wants to write poetry first, finds he can't and then tries the short story which is the most demanding form after poetry. And failing at that, only then does he take up novel writing." Physicists, by comparison, are not so modest. Nevertheless, for a physicist every increase in knowledge of physics is always guided in two ways, by mathematical intuition and empirical observation. The art of physics is to design experiments in order to derive the laws of nature. In this process mathematical intuition is indispensable. In fact, it is easy for mathematicians to switch to studying physics, computer science, or economics, just as it is for poets to turn to writing novels, essays, or plays. Of course, there are exceptions.

Mathematics is usually seen as the diametric opposite of poetry, although there are exceptions here too. Although the opposition is not always true, it stands there basically undeniable. Mathematicians work to discover, while poets work to create. The painter Degas occasionally wrote sonnets and once complained to the poet Mallarmé. He said that he had many ideas—in fact too many, he found it difficult to write. Mallarmé replied, "poems are made not with ideas but with words." On the other hand, mathematicians work mainly on concepts, combining concepts of the same kind. In other words, mathematicians think in an abstract way, while poets think in a concrete way. But again this is not always the case.

Both mathematics and poetry are products of imagination. For a pure mathematician, his or her materials are like lacework, leaves on a tree, a patch of grass, or the light and shade on a person's face. In other words, "inspiration," which Plato denounced as "a mania of poets," is equally important to mathematicians. For example, Goethe fancied that he saw a flash of light when he heard of his friend Jerusalem's suicide. He immediately came up with the outline of *The Sorrows of Young Werther*. He recalled that he "seemed to have written the book unconsciously." Another example: Gauss, "the prince of mathematics," wrote to tell a friend after solving a problem (symbols of Gaussian summation) which had been bothering him for years, "Finally, 2 days ago, I succeeded—not on account of my hard efforts, but by the grace of the Lord. Like a sudden flash of lightning, the riddle was solved. I am unable to say what the conducting thread was that connected what I previously knew with what made my success possible."

Mathematics often appears to be connected to and interactive with astronomy, physics, and other branches of natural science, but it is a completely self-referential and vast field of knowledge with a reality more enduring than other sciences. It is like a true language, which not only records and expresses ideas and the process of thinking but also creates itself through poets and

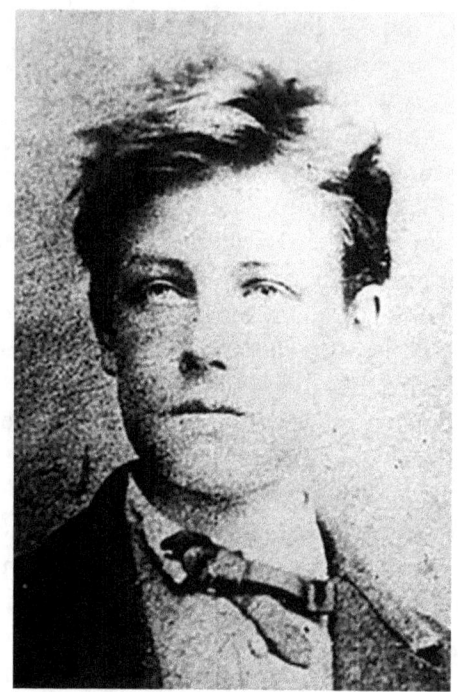

Poet Arthur Rimbaud

Mathematician Évariste Galois

writers. It could be said that mathematics and poetry are the freest intellectual activities of human beings. The Hungarian mathematician Paul Turàn maintained that "Our mathematics is a strong fortress." His words correspond to Faulkner's "People will never be destroyed as long as they yearn for freedom," when he talks about creative writing.

Through years of study and practice, I have come to believe that the process of mathematical research is more or less an exercise or an appreciation of intelligence. This is perhaps one of the main reasons for its great charm. I fully understand what the philosopher George Santayana said in his later years, "If my teachers had begun by telling me that mathematics was pure play with presuppositions, and wholly in the air, I might have become a good mathematician, because I am happy enough in the realm of essence." Of course, I cannot rule out the possibility that a great thinker can yield to the intellectual fashions of his times as a man or a woman can do to fashions in dress.

Compared with other disciplines, mathematics is often an undertaking for the younger. The Fields Medal, the most renowned mathematical prize, goes only to mathematicians under 40. Riemann died at 40, Pascal at 39, Ramanujan at 33, Eisenstein at 29, Abel at 27, and Galois at 20; by the time they died, they had all left their deep traces on the history of mathematics. Some mathematicians, such as Newton and Gauss, lived long lives, but they completed their major work in their youth. Of course, there are exceptions here too.

Likewise, we can draw up a long list of poets who died young: Pushkin, Lorca, and Apollinaire died at 38, Rimbaud at 37, Wilde at 34, Mayakovsky at 22, Plath at 31, Shelley and Yesenin at 30, Novalis at 29, Keats and Petofi at 26,[1] and Lautréamont at 24. Whereas if we look at painting, Gauguin, Rousseau, and Kandinsky began their artistic careers after they turned 30. Thus, more often than other servants of creation, poets and mathematicians tend to burn up the flower of their talent in the midst of the youth. Poets may destroy the shapes common to the forms of their predecessors, in order to renew the form and language; mathematicians may be, by the nature of their industry, more prone to continuity. Again, there are exceptions.

The language of poets is renowned for its conciseness. Ezra Pound is praised as a master of the concise; no one seems to do better than him in this regard. But the language of mathematicians is also noted for its conciseness. The British writer Jerome K. Jerome gave an example, as follows:

---

[1] The Hungarian poet Petofi disappeared in a battle against the Russian-Austria alliance in 1849. He was considered to "have died at the points of the lances of Cossack soldiers" until the end of the nineteenth century, when Russian researchers found in archives that he had actually been taken to Siberia as a prisoner of war and died there of tuberculosis in 1856. He would therefore have been 33 when he died.

When a twelfth-century youth fell in love he did not take three paces backward, gaze into her eyes, and tell her she was too beautiful to live. And if, when he got out, he met a man and broke his head—the other man's head, I mean—then that proved that his—the first fellow's—girl was a pretty girl. But if the other fellow broke his head—not his own, you know, but the other fellow's—the other fellow to the second fellow, that is…

As he goes on to say, this interminable paragraph would be very succinct if expressed in mathematical symbols, although it would be less amusing:

If A broke B's head, then A's girl was a pretty girl; but if B broke A's head, then A's girl wasn't a pretty girl, but B's girl was.

Of course, it would have been less amusing. The language of mathematicians is universal. Goethe joked that mathematicians are like the French, who can translate whatever you say into their own language and turn it immediately into something totally new. We have been taught that a branch of science is truly developed only when it is able to make use of mathematics. In the same way, poetry is a common key factor of all the arts. It can be said that every work of art needs "poetic flavor." Mozart had a reputation as "the poet of music" and Chopin "the poet of the piano." It's not difficult to imagine the striking symmetry between a beautiful mathematical formula in a scientific paper and several brilliant lines of poetry in an essay or a speech.

Now let's come back to the proposition stated at the beginning of this essay. Freud said, "Everywhere I go, I find that a poet has been there before me." This remark was taken up by Breton, the leader of surrealism, as a golden rule. Novalis asserted, "Poetry is very similar to prophecy in its significance. Generally, poems are like the intuitions of prophets. Poets—prophets—reveal the secrets of a strange and wondrous world with magic lines and images." Therefore, a poet of integrity will inevitably violate the interests of those in power. Plato accused poets of being the enemies of truth and their poetry of spreading mental poison.[2] On the other hand, pure mathematics, especially modern mathematics, often develops in advance of its time, even in advance of theoretical physics. It was more than a full century after the invention of Galois's group theory and Hamilton's theory of quaternions that these theories were applied to quantum mechanics. In similar situations, non-Euclidean geometry was used to describe gravitational fields and complex analysis to

---

[2] Plato was always precise in his diction. In his last work, he described those who ignored the importance of mathematics in the pursuing of ideals as "piggish."

Statue of Saint Augustine

describe electrodynamics. The discovery of conic section, which for over 2000 years was considered no more than "the unprofitable amusement of a speculative brain," ultimately found its application raised from in Newton's equation of motion, theory of projectile motion, and the law of universal gravitation.

However, more often than not, the work that mathematicians do is not understood by the crowd. Some people have rebuked them for indulging in pointless speculation or being silly and useless dreamers. Lamentably, this viewpoint of these learned scholars is still supported by certain authorities. For example, Schopenhauer, a distinguished modern philosopher, acknowledged poetry as the highest art but described arithmetic as the lowest activity of the spirit.[3] Since the beginning of the twentieth century, more and more

---

[3] This viewpoint of Schopenhauer is completely contrary to that of Plato, who proclaimed that he would drive poets out of his ideal city and that "God is a geometrician."

Chilean poet and physicist Nicanor Parra

French poet and mathematician Jacques Roubaud

people have come to realize how our times have benefited from mathematics. To some extent, however, poets and artists are still in the situation they always have been. Perhaps they should console themselves with Picasso's words: "People earn the title of artists only after they have overcome innumerable obstacles. Therefore art should be restricted instead of being encouraged."

9 Mathematicians and Poets

By coincidence, mathematicians and poets often walk side by side on the frontiers of human civilization. Euclid's *Elements* and Aristotle's *Poetics*, the two most important academic works of ancient Greece, were written at almost the same time. They both had what one might call a common belief or attitude consisting, one might say, in an accurate "imitation" of the outer world. The difference is that the former was an abstract imitation while the latter was a concrete one. Allan Poe and Baudelaire, pioneers of modern art, belonged to the same age as Lobachevsky and Bolyai, founders of non-Euclidean geometry. When a group of poets and painters of great talent gathered in Paris, in the 1930s and 1940s, to launch the radical revolution of surrealism, some other brilliant minds in the world were working hard in their own way to develop topology, a burgeoning branch of mathematics. Here I want to quote an example, often cited by topologists, which uses a parody of *The Song of Hiawatha* by the American poet Longfellow. It tells of a Native America who made fur mittens:

He, to get the warm side inside, Put the inside (skin side) outside; He, to get the cold side outside, Put the warm side (fur side) inside…

Interestingly, the Native American is actually performing a topological action. The word topology first appeared as *Topologie* in German, in the work of a student of Gauss in 1847, when the concept was known to very few mathematicians.

Finally, I'm going to raise the question of whether someone can be a poet and a mathematician at the same time. Pascal assures us at the beginning of his *Pensées*: "As long as geometricians have good insight, they can be sensitive; as long as sensitive people can apply their insight to geometric principles, they can be geometricians too." Despite this, historically only the eighteenth-century Italian mathematician Mascheroni and the nineteenth-century French mathematician Cauchy could possibly be counted as poets, while the twentieth-century Chilean poet Parra was a professor of mathematics. Perhaps the only one in human history who made great contributions in both fields was Omar Khayyam, the eleventh-century Persian who was born four centuries earlier than the versatile Da Vinci. He made his mark in the history of mathematics for his geometric solution of cubic equations; and he became known to the world as the author of the *Rubáiyát*. When the 14-year-old T.S. Eliot came across Edward Fitzgerald's English translation of the *Rubáiyát* at the turn of the twentieth century, he immediately became enthralled. He recalled the splendor of entering the world of this magnificent poem and realized, after reading those lines full of "dazzling, sweet, and painful colors," that he wanted to be a poet.

(Translated by Robert Berold and Gu Ye)

Acknowledgments The author is grateful to the referees of the Notice of AMS for their valuable opinions and suggestions, particularly to Prof. Preda Mihailescu for insightful, precious ideas and discussions offered during my visit at the Mathematisches Institut, Universität Göttingen. They all made this paper more readable.

10

Mathematicians and Political Leaders

Mathematics, like politics, is the art of possibility.

Mathematicians seldom concern themselves with politics. Nor do they enjoy stirring up trouble the way artists do. Baudelaire, the French poet sometimes called "the father of modern literature," spent his whole life pursuing inspiration in noble ladies' boudoirs. His unrestrained bohemian life ended in misery before the publication of *Paris Spleen*, his posthumous collection of prose poems. In this book he quotes Pascal, the seventeenth-century French mathematician and thinker, as saying: "Almost all our evils arise from being unable to stay in our rooms." Probably because they are more inclined to remain in their rooms, mathematicians are more likely than artists to win the trust and friendship of kings and political rulers.

In ancient Greece, Euclid was a master of geometry. Although his birthplace and exact dates remain unknown, we do know that he studied at Plato's Academy in Athens and later headed the Department of Mathematics at Alexandria University on the invitation of the Egyptian King Ptolemy. There, in its well-stocked library, he completed his famous *Elements*. As a foundation stone of modern science, this masterpiece established an excellent model for deductive reasoning, even enlightening thinkers and philosophers. Allegedly Ptolemy asked Euclid whether there was a shortcut for learning geometry: he was told that there was no royal road. When a student wanted to know the benefits of learning geometry, Euclid ordered a slave to give him a penny and said to onlookers: "This is because he always wants to profit from learning."

Statue of Euclid

Archimedes, born in the generation after Euclid, was the greatest mathematician and scientist of the ancient world. When he was young, he too studied at Alexandria University with Euclid's disciples. When Archimedes returned to his hometown Seracusa, he was highly regarded by King Hiero. According to a widespread story, King Hiero had received a golden crown but was worried that the gold might have silver in it. He sought advice from

10 Mathematicians and Political Leaders

Statue of Claudius

Archimedes. Archimedes had noticed when taking a bath that water was displaced equal to the volume of his body. He immediately realized that an object of low density displaced more water than an object of the same weight having a higher density. Thus, he invented the law of floating bodies and solved the king's problem.

The emperor Claudius in the first-century AD was the first Roman ruler to expand his territory to North Africa. He also sent the Roman army across the English Channel and made Britain a province of his empire. Apart from his military talents, he studied history and wrote voluminous books on history in Greek. He made a foray into mathematics with a pamphlet entitled *How to Win at Dice*, in which he discussed probability. It turned out that he was obsessed with dice and loved gambling with his idle ministers. Unfortunately

Likeness of Pope Sylvester II on a French Stamp

Sultan and scholar Ulugh Beg

10 Mathematicians and Political Leaders

the book has not survived. When Pascal and Fermat laid down the foundations of probability theory in their correspondence in 1654, their starting point was games of gambling such as dice.

The Middle Ages, also known as the dark ages, were not a bad time for mathematicians. Pope Sylvester II had a strong attraction to math. There is evidence that he introduced Arabic numerals, including zero, to Europe. He was said to have devised the abacus, the globe, and the clock. He wrote a book titled *Geometry* in which he solved a daunting task for his contemporaries: Given the hypotenuse and area of a right angled triangle, what are the lengths of the two other sides? Sylvester II, whose real name was Gerbert, had been born, like Claudius, in France. When he was young, he sojourned in Spain where he learned the "Four Arts" in a monastery which attributed its high mathematical standard to Spain's period of Arab rule. Later he went to Rome, where his mathematical talent was noticed by the pope. He was introduced to the Roman emperor, who hired him as tutor to the crown prince. He was regarded highly by succeeding emperors until eventually he was appointed pope himself.

The most outstanding mathematician of the Middle Ages was Fibonacci, known as "Leonardo of Pisa" to distinguish him from da Vinci, "Leonardo of Florence." His talent attracted the attention of Frederick II, King of Sicily, who invited him to court. He successfully answered three mathematical problems put to him by the king's courtiers, and the king and his successor became Fibonacci's patrons. Fibonacci's "rabbit problem" remains a mathematical riddle even today. A mathematical journal called *The Fibonacci Quarterly* published from the barren expanses of South Dakota is devoted exclusively to the "rabbit problem." There is also a large Fibonacci Association which holds biannual conferences around the world.

In the East, Fibonacci's contemporary Li Ye, a Chinese mathematician in the Yuan Dynasty, was summoned three times to the court of Kublai Khan. This was not so much because the Khan needed mathematical education, but because being from outside China, he wanted to win over Chinese intellectuals. What "the Occupier" valued was not Li Ye's mathematical talents but his capacity as a successful candidate in the highest imperial examination, especially his reputation as a learned scholar.

In the Arab world and Persia, monarchs attached great importance to mathematics. In the ninth century, Abbasid caliph Al-Mamum ordered the construction of House of Wisdom in the capital Baghdad. As a combination of library, scientific academy, and translation agency, it became the world's most important academic institution after the Library of Alexandria. Al-Khwarizmi, the Baptist of Algebra, was hired as its director.

Arab-occupied Persia had some outstanding mathematicians, almost all of them protected and sponsored by monarchs. For example, the eleventh-century Seljukian sultan Malekshah invited Khayyam to the capital Esfahan to lead work on building an observatory and compiling the calendar. Khayyam spent most of his life there until the sultan died. Their story was made into the Hollywood movie *Omar Khayyam* (1957). Similar patron-mathematician relationships included the thirteenth-century mathematician Nasir al-Din with Hulagu Khan and the fifteenth-century example of al-Kashi and the scholarly Timurid prince Ulugh Beg. The luckiest was al-Kashi, who died before the prince. Had the prince died before al-Kashi, the mathematician might have been persecuted by his patron's political opponents, just as Omar Khayyam had been. As Ulugh Beg wrote, al-Kashi was "a remarkable scientist, one of the most famous in the world, who had a perfect command of the science of the ancients, who contributed to its development, and who could solve the most difficult problems." Al-Kashi determined Pi to 17 decimal places, breaking the 900-year-plus record of Chinese mathematician Zu Chongzhi.

In modern European history, some enlightened monarchs were close friends of mathematicians. The seventeenth-century Swedish queen Christina invited the French mathematician and philosopher Descartes to her court, eventually sending a warship to the Netherlands to welcome him. Descartes, who was frail and lived in seclusion, was hesitant and decided to go only at the last moment, won over by the queen's enthusiasm and sincerity. His worries turned out to be well justified. The cold air of Stockholm brought him pneumonia, and he died there 4 months later. The 1933 Hollywood movie *Queen Christina of Sweden* starring Greta Garbo relates the story well.

In the eighteenth century, the Swiss mathematician Euler was twice hired by the St. Petersburg Academy. He spent 31 years at the academy, where his teachers were two members of the distinguished mathematical Bernoulli family. Euler, who became blind during his long stay in Russia, is among the most prolific and respected mathematicians in history. During the 25 years that he headed the Prussian Academy in Berlin at the invitation of Frederick the Great, St Petersburg continued paying his salary. When Euler returned to St Petersburg, Frederick the Great extended an enthusiastic invitation to Lagrange, the Italian mathematician who had settled in Paris. "The greatest king of Europe" wished "the greatest mathematician of Europe" could stay in his court. Obviously the king took Euler's departure to heart.

Of all the European rulers, Napoleon shared the closest relationship with mathematicians, befriending almost every important French mathematician of the time. After his expedition to Egypt, Napoleon declared Lagrange to be

10 Mathematicians and Political Leaders

Napoleon the amateur geometer

the towering pyramid of mathematical science. Once he asked Laplace jokingly, "why didn't you mention God in your work?", the mathematician answered, "I had no need of that hypothesis." However, Lagrange, Laplace, and another L—Legendre—all avoided the French Revolution. Napoleon was a good geometer himself. He raised the question of dividing a circle into four equal arcs using only a compass. This problem was later solved by his friend Mascheroni, an Italian mathematician who had also settled in France.

When Napoleon's army retreated from Moscow in 1812, hundreds of thousands of French combatants were taken as prisoners of war. One of these was Poncelet, a 24-year-old mathematician. In the prison camp he conceived his masterpiece *Treatise on the Projective Properties of Figures*. Back home after he was released, he published the book in Paris in 1822, inaugurating a "glorious age" in the history of projective geometry. But Napoleon unwittingly hurt a great mathematician, Gauss, the "Prince of Mathematics." Born into an ordinary working family, the precocious child attracted the attention of Duke Ferdinand from his hometown in Brunswick, Germany. The Duke became Gauss's patron and close friend. Although he died anonymously when Gauss was 29 during the invasion of Napoleon's army, Duke Ferdinand had won immortal fame in the history of mathematics.

Across the Atlantic Ocean, a number of American presidents were closely related to mathematics. George Washington was said to be a competent surveyor. Thomas Jefferson worked at encouraging the teaching of advanced mathematics. Abraham Lincoln tried to teach himself logical thinking by studying Euclid's *Elements*. The most mathematically gifted of them all was James A. Garfield, the 20th president. Although his political performance was average and he was assassinated while in office, he demonstrated outstanding ability. During a mathematical discussion with some members of Congress in 1876, he worked out a very simple proof of the Pythagorean theorem. The proof depends on comparing and simplifying two different ways of calculating the area of a trapezoid (the first way is by using the area formula of a trapezoid, the second by summing up the areas of the three right angled triangles that can be constructed in the trapezoid). Compared with da Vinci's proof of 400 years earlier, Garfield's solution was far more elegant. When I was in Washington, I wondered whether it was for this reason that his bronze statue had been placed in front of Capitol Hill, an honor accorded to nobody else.

Back across the Atlantic we find Isaac Newton, praised with Archimedes and Gauss as one of the three greatest mathematicians in history. His major mathematical achievement was the discovery of calculus, but he is also generally attributed with discovering the law of gravity and the laws of motion, all of which are expressed by elegant mathematical formulae. Newton's mathematical achievements, together with his prominent contribution in physics and astronomy, sent him into Parliament on behalf of his university, and in time he became the first scientist to be knighted by Queen Anne. But the only record of Newton's parliamentary utterances was a request to open a window, and in his later years he indulged in theology. However, he was promoted to the powerful post of Warden of the Royal Mint.

Leibniz, born in Leipzig, had a more outgoing personality. He enjoyed socializing with aristocrats from an early age. In his time Germany had not yet been unified and had fallen behind in science, technology, and military power,

10 Mathematicians and Political Leaders

George Washington, land surveyor

making it vulnerable to annexation by other powers like France. In 1672, the distressed Elector of Mainz sent the eloquent Leibniz to Paris. His mission was to distract Louis XIV away from any predatory interest in Northern Europe and entice him with a scheme to rather conquer Egypt. Leibniz ended up not seeing the French king at all, but he stayed on in Paris to study mathematics. He invented calculus on his own, sparking off a priority dispute with Newton which impeded mathematical development in Britain for a century.

Some mathematicians did engage more actively in politics. The first great mathematician in ancient Greece, Pythagoras, was one. He and his disciples organized an association in Crotone in the southern Apennine Peninsula, allying with aristocrats and being driven away by democratic parties. Pythagoras fled to Metapontum nearby, where he was killed in 497 BC.

Archimedes was stabbed to death by a Roman soldier during the invasion of Syracuse, although his murder does not seem to have been linked to his close relationship with King Hiero. Years later, when the Roman statesman and writer Cicero visited the island of Sicily, it is said that nobody was willing to tell him where Archimedes's grave was to be found. The great orator had to clear away thorns and search it for himself.

In France, the father of differential geometry, Monge, was ridiculed for closely following Napoleon through Napoleon's long career. He accompanied Napoleon on the expedition to Egypt, along with Fourier, inventor of trigonometric series. When they came back, Monge became a government minister while Fourier was made only a prefect. Monge's student Lazare Carnot, an enthusiastic revolutionary and a fine military strategist, was praised as the "organizer of victory." But Carnot was the only member of the Committee of Public Safety who dared oppose Napoleon's crowning. He was forced to flee to Geneva and died abroad in hunger and poverty. Since Carnot was so deeply involved in politics, his scientific work was relegated to a spare time hobby. His offspring seem to have inherited his abilities in separate packages. One of his sons became Minister of Education; another was a prominent physicist. One of his grandsons became president of France; another was a reputable chemist.

Laplace, known as "the French Newton," fared better in the political world than his contemporary, Carnot. He was born into a Normandy farmer's family, but his talent soon landed him important positions under Louis XVI. He displayed a remarkable ability to survive political turbulence. During the French Revolution, he received a special pardon in exchange for calculating the trajectory of artillery shells. Soon after the revolution, he became politically active again under Napoleon—he had once been Napoleon's math teacher. He took charge of the Bureau of Longitude, was a temporary Minister of the Interior for 6 weeks until Napoleon's younger brother replaced him, and then became Chancellor of the Senate. After the Bourbon Restoration, he moved with the times, swore allegiance to Louis XVIII, and was given the rank of Marquis.

However prominent political leaders can be while in office, they are easily forgotten when they resign or die. In an article written for the 300th anniversary of Newton's masterpiece *Philosophiae Naturalis Principia Mathematica*, the British scholar R S Westfall remarked that we never celebrate the 300th anniversary of a civil official. Perhaps that's true for the United Kingdom, but there have been some great monarchs in history whose memory has endured— Alexander the Great, Augustine, Genghis Khan, and Ashoka come to mind. Unlike politicians, some great mathematicians tend to enjoy broad and lasting appeal, which probably derives from the durability of mathematics itself.

The fifth-century Byzantine scholar Proclus, often considered the last major Greek philosopher, was head of Plato's Academy in Athens in his later years. He suggested why mathematics had enduring qualities:

> Mathematics is something that reminds you of your invisible soul, bestows life on the truth that it discovers, awakens the mind, clarifies wisdom, sheds light on our inner thoughts and washes away our inherent ignorance.

10 Mathematicians and Political Leaders

Sir Isaac Newton Bar (photograph by the author, Cambridge)

Laplace Station (photograph by the author, Paris)

The seventh-century Arabic astronomer and mathematician Brahmagupta wrote something similar, in poetic language:

> As the sun eclipses the stars by its brilliance, so the man of knowledge will eclipse the fame of others in assemblies of the people if he proposes algebraic problems, and still more if he solves them.

As far as I am concerned, the extensive application of mathematics in today's world has made being a mathematician a desirable occupation for modern people. Meanwhile, mathematics has the advantage of offering a spiritual and intellectual refuge from the hustle and bustle of the world. In a sense, both mathematics and politics are arts of probability. People who practice them need enterprise and courage. A great mathematician or political leader relies on intuition and luck when facing complicated problems, although he or she tends to have limited experience and wisdom outside mathematics or politics. Such limitations alienate them from the masses. Despite this, both mathematicians and political leaders have their unique spirituality and way of life.

What, after all, is greatness? Pascal classifies several types of greatness in his *Pensées*. The first type is physical: material greatness represented by the visibility of space and stars and the material trappings of kings and powerful or wealthy rulers. The second type is the spiritual and intellectual greatness of geniuses such as Archimedes, Newton, and Gauss. They have their own kind of field, power, glory, and luster, which is perceived not by the eye but by the mind. As Pascal further pointed out, "all the glory of greatness has no luster for those in search of understanding."

Hangzhou 2000, Cambridge 2008

(Translated by Robert Berold and Gu Ye)

11

Hua Luogeng and Shiing-Shen Chern: Two Contemporary Chinese Masters

In these days the angel of topology and the devil of abstract algebra fight for the soul of each individual mathematical domain.
—Hermann Weyl

Northwest and Southeast of Tai Lake

At the end of the nineteenth century and the start of the twentieth, the Tai Lake basin in eastern China was brimming with talent, and many masterful figures were born there, much as had been the case in the Poyang Lake basin during the Song dynasty. It is no exaggeration to say that more than half of the literary and scientific giants of modern China came from this region, which today we accustomed to referring to as the Yangtze River Delta, although this mainly reflects an economic perspective, associating it with the Pearl River Delta, the first frontier of reform and opening-up. From the perspective of historical and cultural origins, there is no question that this region is more closely related to Tai Lake than to the Yangtze River. On the northern and southern shores of the lake, respectively, there are the six cities of the Suzhou-Wuxi-Changzhou metropolitan area in Jiangsu province and Hangzhou, Jiaxing, and Huzhou in Zhejiang. These six cities belonged to the Liangzhe Circuit of the Song dynasty, representing the *land of fish and rice*

Throughout this chapter, I have used the standard (pinyin) romanization of all Chinese names except in case the individual in question is already better known by a different romanization in English, as, for example, Shiing-Shen Chern instead of Chen Xingshen. (Translator's Note)

Young Hua Luogeng

Young Shiing-Shen Chern

popular among the Chinese people and also the beautiful Jiangnan held in esteem in the poems of the literati.

The mathematical prodigy Hua Luogeng was born on November 12, 1910, into a family of small business in Jintan county in the city of Changzhou. His father had been an apprentice, and years of hard work had enabled him to own three shops of various sizes and to serve at one time as trustee of the county Commercial Silk Association. The largest of these stores was unfortunately destroyed in a fire, and later the next largest also shut its doors, so that, by the time Hua was born, there remained only the one small cotton shop. Jintan (meaning *golden altar*) is to the northwest of Lake Tai. In Xiushui (which means *beautiful water* and is now called Jiaxing) to its southeast, another extraordinary genius was born less than a year later, who was destined to become Hua's roommate, colleague, and competitor in the future. This was Shiing-Shen Chern, whose family background was entirely different. His father was a scholar who had obtained the position of *xiucai* in the imperial examination system and who was admitted to the school of law and politics in the provincial capital, Hangzhou, after the birth of his son.

After he had graduated, Chern's father worked on judicial matters and rarely returned home; Chern learned to read and developed his literacy with his loving grandmother and young aunt. Once when his father had come back to Jiaxing to celebrate the Chinese New Year, he taught his son Arabic numerals and the four operations of arithmetic; he also left behind a collection on handwritten mathematics compiled by missionaries. Surprisingly, Chern could solve most of the exercises in this book even at such a young age, igniting his interest in mathematics. Due to the doting nature of his family, Chern attended elementary school for only a single day before he dropped out and returned home. When he was a little older, he attended the upper primary school of Xiuzhou Middle School, where he was not only able to tackle quite complex mathematical problems, but also took a liking to Chinese, and was able to read such books as the *Investiture of the Gods*,[1] cultivating his literary predilections. He even published two free verse poems in the school magazine. Chern was still in his hometown in the summer of 1921, when Zhang Guotao, Mao Zedong, and others who participated in the First National Congress of the Chinese Communist Party changed locations in secret from Shanghai to a tourist boat on South Lake in Jiaxing. But the following year, his father was transferred to the court of Tianjin, and the family left Jiaxing.

In the same year, Hua had entered Jintan Junior Middle School. During elementary school, his grades had been less than stellar due to naughtiness, and he obtained only a certificate of education for his time there; but it was his sister who was compelled to abandon school in spite of her better grades because their father regarded men as superior to women. From the second year on, however, the school mathematics teacher began to look at Hua in a different light, and by the third grade Hua had taken great steps to simplify the solutions in his exercise book; he also made progress in Chinese language and took an interest in ancient poetry. When Hua Luogeng graduated from middle school, his father found himself again in some trouble. On the one hand he hoped for his son to succeed in his studies and pursue the path toward becoming an official; on the other, he worried that sending his son off to a high school in the provincial capital would be too financially burdensome. In the end, Hua matriculated to a vocational school administered by

---

[1] A Ming dynasty vernacular novel of mythology and fantasy (Translator's Note).

the educator Huang Yanpei in Shanghai; here, there was no tuition fee and Hua was obliged only to pay for his room, board, and miscellaneous fees. He was admitted to the business program of the school, equivalent to modern day junior high school.

This was in 1926. In Tianjin, Chern, who was 1 year younger, graduated from Rotary Middle School (now Tianjin Railway No. 1 Middle School), where Zhan Tianyou[2] had served as a trustee, skipped entirely over college preparatory programs, and entered directly into Nankai University. Meanwhile, Hua had won first place in the Shanghai Abacus Competition, but he was forced to drop out of school due to the financial difficulties of his family and returned home to help his father man the sales counter. At the age of 16, Hua married a girl surnamed Wu from the same city. Chern, by contrast, was 28 by the time he married, and by that time, he had already earned a doctorate abroad and obtained appointment as a university professor. It is worth mentioning that in his youth Hua had a tall figure and upright appearance (his daughter, Hua Su, personally told the author that his was 1.8 m tall). His wife gave birth to a daughter in the second year of their marriage. But Hua still loved to read mathematical books and work out exercises in calculation, sometimes becoming so absorbed in thought that he would forget to receive customers entirely, much to the chagrin of his father.

After another year had passed, a former junior high school teacher named Wang Weike, who had recognized and admired the studious character of Hua, returned from his studies at the University of Paris and took over as the principal of Jintan County Elementary Secondary School. Considering the difficult financial straits of the Hua family, he hired Hua to work as the school accountant. Principal Wang was not only a natural scientist, who had attended the classes of Marie Curie, but also an accomplished translator: he was the first Chinese translator of both the Italian poet Dante Alighieri's *Divine Comedy* and the Indian epic *Sakuntala*. Middle school teachers at that time were not only highly knowledgeable but also motivated by a sincere love for their students. The previous principal, Han Damou, also published many books, such as *Introduction to Exegesis*, and taught his students in a step by step fashion, both in life and in learning. Although Hua is regarded now as the model of a self-taught talent, in fact he benefited quite a bit from his junior high school, not only in terms of his knowledge; but this is missing in the education system of today.

Just as Principal Wang was preparing to promote Hua to the position of mathematics teacher, however, a series of misfortunes befell the Hua family

---

[2] The "father of China's railroads".

one after another. First, his mother died of uterine cancer, and then Hua himself fell sick with typhoid fever, which left him bedridden for half a year. Although he survived, the ordeal left him disabled: while walking, he had to trace a circle with his left leg before his right leg could catch up to it, for which he earned the comic nickname *ruler and compass*. But it was precisely this disability that strengthened his resolve to study mathematics, and otherwise he may very well have turned his talents in another direction. In December of that year, the Shanghai periodical *Science* published in the form of a letter from a reader a note that he had written showing that a recent paper claiming to have solved the quintic equation was fatally flawed. This note changed the trajectory of his fate.

As for the general solution to the quintic equation, it had been a difficult problem and a focal point of mathematical attention since the sixteenth century, when several Italian mathematicians produced general solutions to the cubic and quartic equations. In 1824, the young Norwegian mathematician Niels Henrik Abel proved that no such general solution exists. But the Shanghai magazine *Xueyi* (*Learning*) had published a paper purporting to provide a solution in 1926. Of course the experts knew at a glance that the result could not be correct, but nobody could point to any concrete error in it. Young and still unknown, Hua read this paper carefully and mulled it over, finally discovering that there was a mistake in the calculation of a 12th-order determinant. He wrote up his reasoning and refutation and became famous in one fell swoop. Xiong Qinglai, head of the mathematics department at Tsinghua University, read Hua's article cheerfully with his colleagues. After some inquiries, they tracked down Hua and extended an invitation.

Hua was still only 20, and Chern, at the age of 19, was about to graduate from Nankai University. Prior to enrolling, Chern had moved further in the direction of mathematics due to the influence of Qian Baocong, a teacher at the school and a historian of mathematics, also from his hometown; in fact, their fathers had been middle school classmates and later studied abroad in the United Kingdom. At the time, the freshmen of the School of Science and Technology at Nankai University were not divided into departments, and on one occasion in a chemistry lesson, Chern was asked by the teacher to blow a glass pipe. He found himself helpless in the faces of the flame and the pieces of glass, and although he managed to blow the pipe, he feared that it was still too hot and ran it under water, causing the tube to shatter immediately. Chern came to the conclusion that his ability for hands-on work was poor, and he decided to give up chemistry and physics and devote himself to mathematics. Chern quickly attracted the admiration of Jiang Lifu, head of the mathematics department, with a Ph.D. from Harvard University, and he became interested in geometry. On the eve of his graduation, Chern was admitted to Tsinghua University as the first graduate student in China (the research institute there had the opportunity to select outstanding students to study abroad at public expense). At Tsinghua, Shiing-Shen Chern would meet Hua Luogeng, who arrived later, and the two of them would turn a new page in the history of Chinese mathematics.

From Tsinghua University to Europe

The scientific foundations in old China were weak, especially prior to 1930. Anybody with a doctorate degree from a foreign country was able to obtain immediate employment as a professor, with good pay and no worries for the necessities of daily life. These professors were overburdened with teaching and lacked access to current information or a good academic atmosphere and environment; they basically gave up on research. Consider Jiang Lifu, for example, who was the only teacher in the Department of Mathematics at Nankai University for the first 4 years of his tenure there and had to teach every class himself in person. In 1949, he founded the Department of Mathematics at Lingnan University in Guangzhou, which later merged with Sun Yat-sen University. Xiong Qinglai had earned only a master's degree in France when he founded and became the first director of the mathematics departments at Southeast University (later renamed Central University) and Tsinghua University.

However, Tsinghua University is a royal university, established first as Tsinghua College and supported with the Boxer Indemnities, returned to China by the United States, which was also used to fund young talents such as Jiang Lifu to study abroad. At the beginning of the last century, the Eight-Nation Alliance, which included Britain and the United States, had brutally suppressed the Boxer Rebellion on the pretext of protecting their own clergy and expatriates, which had broader impacts on China at the time, including the rules for the return and use of the Boxer Indemnities. Otherwise, the Qing dynasty would probably not have been willing to allocate so much money all at once for the development of education and the support of selected outstanding scholars for study abroad. Many of these young people later became pillars of the country and are well known to us today.

While it was still Tsinghua College, the school invited Zheng Tongsun, who had earned a master's degree from Cornell University and later became father-in-law to Chern to serve as the director of the undergraduate program of the Department of Mathematics. It was at his recommendation that Xiong

Tsinghua College

Qinglai became the director of the mathematics department at the renamed Tsinghua University in 1928. Sun Guangyuan and Yang Wuzhi (father of Chen-Ning Yang), with doctorates from the University of Chicago, joined shortly thereafter. Of these four, however, only Sun Guangyuan continued to carry out research, mainly in differential geometry. Sun Guangyuan was a native of Hangzhou in Zhejiang Province, so he partially shared his hometown with Chern, who became his graduate student at Tsinghua University. That year, the university admitted two graduate students, but the second, Wu Daren, postponed his admissions for family reasons. It was decided that Chern should be permitted to serve as a teaching assistant for 1 year.

In August of the following year, just as Chern was beginning his graduate studies, Hua arrived at Tsinghua University. As an assistant, his office was just outside that of the department head, Xiong Qinglai, so that anybody coming to visit the latter would run into Hua. He was outgoing and funny and quickly became familiar with everyone, including Chern. His salary at that time was only half that of a teaching assistant, slightly higher than that of a laborer, and similar to the graduate student allowance afforded to Chern. His family had stayed behind in Jintan, and his wife gave birth to a son that year. Due to financial difficulties, Hua was only able to return to his hometown for winter and summer vacations during his 5 years at Tsinghua University. In his biography *Hua Luogeng*, Wang Yuan recorded a sweet recollection of this period from the later years of his mentor: "Whenever I went back to my hometown to visit relatives during the winter and summer vacations, Mr. Xiong Qinglai was always reluctant to see me go. He was afraid that I would not come back, because I earned too little money. He had no way of knowing that the salary given to me by Tsinghua University was much more than that given to me by Jintan Middle School. Tsinghua University was a dream for me."

Although Hua was able to publish four papers in *Science* in the year he came to Tsinghua University, on the strength of his legendary previous work, these papers were all written in his hometown and involved only elementary mathematics. After he arrived, he eagerly attended lectures and studied advanced mathematics, and did not publish any papers for the next 2 years. Chern later wrote that this was the most important and successful period of his self-study, in which he completed the university coursework, and began to write papers. From 1934, Hua exhibited greatly improved mathematical potential. Exerting himself to the fullest, he subsequently published six to eight papers annually, mostly in foreign journal, including the German journal *Mathematische Annalen*, which quickly earned him a reputation. Following the lead of Yang Wuzhi, these papers mostly concerned number theory, but some were in algebra and analysis, reflecting the diversity of his interests and talents. He also pursued a unique course of self-study in English.

Just as Hua was beginning to show the fruits of his talent, Chern, who had had lofty goals since his childhood, had passed the defense of his master's thesis and was preparing to study abroad. In July of 1934, the Professorial Council of Tsinghua University passed a motion to send him to Germany to study, with funding still provided by the Boxer Indemnities. Among the professors in attendance at this meeting were his future father-in-law Zheng Tongsun, "matchmaker" Yang Wuzhi, principal Mei Yiqi, litterateur Zhu Ziqing, and so on. At the end of the month, Chern boarded a boat from Shanghai to Europe, heading for the University of Hamburg to study geometry under professor Wilhelm Blaschke, with whom Chern was acquainted via Peking University, in the same city as Tsinghua University.

Whereas Tsinghua University, with its abundant financial resources, was busy building facilities and recruiting talent, Peking University, in spite of its long history, was disorganized and lax in discipline and often in arrears when it came to the salaries of its professors. This situation changed only after Hu Shi, dean of the School of Liberal Arts, became director of the China Education and Culture Foundation, which was in charge of the Boxer Indemnities, and urged it to pass special funding measures for Peking University. The Peking University set up operations 2 years after the establishment of the Tsinghua Research Institute and began to invite foreign experts to present lectures. Among the first mathematicians to come and speak was Professor Blaschke, who presented a lecture series on topological problems in differential geometry. Chern attended every one of these lectures without fail and got to know the great mathematician.

The Elbe River and the River Cam

Hamburg is among the famous cities of Germany and its most important water transportation hub. Massive ships can reach this city from the Atlantic Ocean directly along the Elbe River. In fact, rivers crisscross the city, and there are more than a thousand bridges. The University of Hamburg, however, is very young. It was founded in 1919, the same year as Nankai University. Many other universities in Germany, where science and culture were already highly developed, have longer histories: for example, the universities of Humboldt, Göttingen, Tübingen, and Heidelberg; most prominent among them was Göttingen, which had become the center of the mathematical world due to the presence of David Hilbert. But the primary consideration for Chern was his advisor; otherwise, he could of course have chosen some prestigious school in England, France, or the United States, as most international students do.

In his later years, Chern would describe his success as half talent and half luck; his first bit of luck can be said to have been meeting Professor Blaschke at Peking University. Chern arrived in Hamburg in the autumn of 1934. Hitler had already come to power, and the so-called Civil Service Law, which stipulated that Jews could not be university professors, had been enacted, first affecting prestigious universities such as the University of Göttingen. The new university in Hamburg, which employed no Jewish professors, continued along its ordinary research path. A second such law was promulgated in 1937, forbidding even those with Jewish spouses from being professors. By this time, Chern had already received his doctorate, and his advisor recommended that he study in Paris under the great mathematician Élie Cartan.

Since his scholarship via the Boxer Indemnities afforded him a high allowance, Chern was full of confidence during these years, able to eat in high-end restaurants, invite Chinese compatriots to dinner, and finally move to Paris to continue his studies at his own expense, albeit with continued support from the foundation. During his time in Hamburg, Chern did not focus on completing papers, but rather on learning and mastering the most modern geometric methods and establishing a wide network of contacts. In addition to Blaschke and Cartan, Chern also entered into exchanges with André Weil, a representative of the French Bourbaki mathematical collective, and Oswald Veblen of Princeton University in the United States. It was like a marathon,

Shiing-Shen Chern's alma mater, the round main building of the University of Hamburg (photograph by the author)

in which it was necessary first to keep pace with the highest echelon, waiting for an opportunity to break through and surpass them. It should be mentioned also that Chern was a sincere spirit and good at making friends. Consider as an example his friendship with Cartan: during the most difficult period of the Second World War, he continuously sent food packages from the United States to his mentor.

In contrast, Hua Luogeng, with a financially difficult background and lacking guidance from his parents and influential teachers, was obliged to rely more on personal struggle and self-study and to work particularly hard. Even after he dropped out of school to work in sales at home, he continued to read late into the night and nevertheless wake up early, earlier even than his neighbor, who ran a tofu shop. So when he was later hired by Tsinghua University, Hua cherished all the more the opportunity to study harder and gain more knowledge. In a short period of time, he published a considerable number of research papers both at home and abroad, a different approach than that of Chern, who came from a well-established family. Then, 3 years after Blaschke had visited Peking University, Tsinghua University also invited two even greater mathematicians: the French mathematician Jacques Hadamard and the American mathematician Norbert Wiener, both of whom stayed in Beijing for a longer period of time.

Hadamard had done pioneering work in many areas of mathematics, and he was particularly outstanding in analytic number theory, providing the first proof of the prime number theorem, a celebrated conjecture due to Carl Friedrich Gauss, the "prince of mathematics." This work was completed toward the end of the nineteenth century; more than half a century later, the discovery of an elementary proof of this theorem led to both a Fields Medal and a Wolf Medal. By the time he came to China, Hadamard was already advanced in age and no longer carrying out academic research at the forefront of mathematics. Wiener, on the other hand, was still young and in his academic prime. The inventor of cybernetics, Wiener was the author of a glorious page in the history of mathematics. He came to greatly admire Hua Luogeng and recommended to him that he go to Cambridge University, where Wiener himself had studied for a time in his youth, to study under his former teacher, G.H. Hardy. Once again, the Boxer Indemnities funded the scholarship.

Hua stopped in Berlin during his travel along the Trans-Siberian Railway via Moscow, and Chern came over from Hamburg to get together. At that time, the Summer Olympics were being held in Berlin, and the two of them went together to watch the games with great interest. In autumn of the same year, Chern left Hamburg to go to London and to Paris, and he also went to Cambridge to visit Hua. Of course, since Chern was relaxed in his approach to learning, he likely did not visit Berlin and Cambridge only to meet Hua but also to satisfy his more playful nature. It should also be mentioned that, according to the archives of the Chinese Culture and Education Foundation, Hua had twice been awarded funds to study at the University of Hamburg prior to his stint in Cambridge, but he failed to make the trip. If he had gone to Hamburg at that time, Hua may have studied the promising new directions in algebraic number theory under Erich Hecke or Emil Artin, in which case the future face of Chinese mathematics might have been significantly different.

At the time that Hua arrived at Cambridge, Hardy was in the United States presenting lectures. He had read the letter of recommendation provided by

Wiener, as well as Hua's thesis, and left behind a letter saying that Hua could obtain a doctorate within 2 years. Hua abandoned his coursework, however, in order to save on time and tuition and focused instead on attending lectures, participating in seminars, and writing papers. Hardy returned to Cambridge 1 year later, without having given Hua any special guidance; again Hua found himself relying on his talent for self-study. During his 2 years in Cambridge, Hua wrote more than ten first-class papers, greatly exceeding the level of his previous work. Wang Yuan remarked that in this time Hua had been reborn and become a mature mathematician. Of course, this cannot be divorced from the influence of the presence of a very strong analytic number theory team at Cambridge that worked closely with the top number theorists of the time, including the Soviet mathematician Ivan Vinogradov. This was another important step for Hua, who established academic contacts with Soviet mathematicians that would prove particularly important for his future research.

After 2 years, Hua made his preparations to return to China. When he came to say his goodbye to Hardy, the eminent mathematician asked him what he had achieved during his time in Cambridge, and Hua presented his results one after another. Somewhat taken aback, Hardy replied that he was in the process of writing a book and would add some of these results to it. This became *An Introduction to the Theory of Numbers* (1938), and Hua may have been the first Chinese mathematician in modern times to see his results cited by famous foreign mathematicians. His main achievement at Cambridge concerned the estimation of trigonometric sums, the Hardy-Littlewood circle method and Waring's problem, the Prouhet-Tarry-Escott problem, and Goldbach's conjecture. At the same time, Hua worked on the draft of what would become his masterpiece—*Additive Theory of Prime Numbers*.

From Kunming to Princeton

In 1937, 1 year before Hua returned to China from the United Kingdom, Chern was getting ready to leave Paris. He had been living in Europe for 3 years by that time, and his alma mater, Tsinghua University, hired him as a professor. He could not have predicted that just 3 days before his departure, the Marco Polo Bridge incident would break out and Japan would occupy Beijing. Although the future was uncertain, Chern ignored the danger. During his time in Hamburg, his old teacher Yang Wuzhi wrote to him to introduce him to the daughter of Zheng Tongsun. In fact, Chern had met Ms. Zheng while he was studying at Tsinghua University, and since he had made a good impression on her, the two began to correspond. For that generation, they were considered boyfriend and girlfriend in name only. Although Chern was eager to rush back, he could not help but first take a boat across the Atlantic to make a pilgrimage to Princeton. It was a hot summer, however, and the mathematicians had all beat a retreat to escape the heat. He traveled across the American continent to California before finally boarding a postal liner and returning to Shanghai.

His trip to the United States left Chern with a favorable impression, and 6 years later he would return and spend most of the rest of his life there. In the meantime, when his ship reached the mouth of the Yangtze River, the shores were in flames, and Shanghai had just been occupied by the Japanese. As a last resort, the ship turned around and headed south to Hong Kong. Chern could not meet with his girlfriend in Shanghai and stayed in Hong Kong for more than a month. Only then did he learn that Tsinghua University, Peking University, and Nankai University had merged to form Changsha Temporary University in Hunan. Chern arrived before the start of school but only stayed in Changsha for 2 months before moving on along with it to Kunming further to the south due to the spread of the war. At the end of the year, Chern became engaged to Ms. Zheng, who was still a second-year student in the biology department at Yenching University at the time, and the two would only marry in Kunming a year and half later.

In the same year that Chern reached Kunming, Hua also returned from England, and he was also hired as a professor at the Southwest Associated University, which had taken the place in Kunming. The two of them were, respectively, 26 and 27 years old. Hua traveled there by way of Hong Kong, Saigon, and Hanoi, and his wife and children had arrived before him. The reunited family lived in the suburbs to stay clear of bombs dropping from Japanese aircrafts. The university was also in the suburbs, but far away from their home, and Hua took a bumpy bullock cart ride every time he went to class. Later, when he had classes he would live at the school and share a room with two other bachelors, including Chern. Shortly after Chern had married, his wife became pregnant and returned to Shanghai to stay with her parents.

Both Hua and Chern made new breakthroughs in their mathematical research during their time at Southwest Associated University. The two shared a room for a year, each with a bed, a desk, and a chair. Although the professors at the university at that time lived in poverty and their working conditions were difficult, their enthusiasm for teaching and research was extremely high. There were also many outstanding students, including Chen-Ning Yang, Deng Jiaxian, Tsung-Dao Lee (Li Zhengdao), and so on. For a period of time, Hua and Chern woke up early in the morning to talk and laugh before immersing themselves in their respective mathematical spaces until late at

Former residence of Hua Luogeng during his time at Southwest Associated University, Kunming

night. Although the two of them never coauthored a paper, they jointly organized the seminar on Lie groups at Southwest Associated University, which was quite advanced at that time.

During his time at Southwest Associated University, Hua carried out his research into number theory primarily in collaboration with Min Sihe, who had earned his doctorate at Oxford University, and also carried out postdoctoral research at Princeton University. During this time, Hua also worked diligently to complete his first monograph, known now in English as the Additive Theory of Prime Numbers. He was already and established leader in his field, but not content to rest on his laurels, Hua carried out innovative research into automorphic forms and matrix geometry. The former remains a significant research topic through to today, while the latter is associated with the work of Cartan, who is the postdoctor supervisor of Chern, to whom Hua even extended his gratitude in the conclusion of one his papers, for having furnished a copy of a preprint paper by Cartan. Hua additionally explored several algebraic problems, including topological automorphisms of finite and symplectic groups, which eventually led him to delve more deeply into classical group theory.

Meanwhile, Chern was also making progress. In his second year after returning to China, he published an article in the American journal *Annals of Mathematics*. This publication, which is jointly sponsored by Princeton University and the Institute for Advanced Study, remains the most important journal in mathematics in the world today. In addition to this standout work on integral geometry in Klein spaces, he also did outstanding work in other fields and appeared twice more in the *Annals of Mathematics* over the next several years. The French mathematician Weil wrote a long article in praise of this work for *Mathematical Reviews*, concluding that it surpassed the original achievements of the Blaschke school. These works paved the way for Chern to enter and establish himself in the United States. It was also during this period that he became interested in the Gauss-Bonnet formula.

In the summer of 1943, Chern was invited to do research in the United States. He stayed at Princeton for 2.5 years and completed the best work of his life, including obtaining an intrinsic proof of the Gauss-Bonnet formula, which marked the advent of a new era for global differential geometry. This was within the first 3 months after arriving to the United States, which shows that he had prepared for it fully already in Kunming. Two years later, on the eve of his return to China after receiving news that his mother had fallen critically ill, Chern proposed the invariant theory now known as Chern characteristic classes. At that time, the Anti-Japanese War had succeeded, and Hua was living like a duck in water in China on the strength of his personal achievements and social skills. He was invited first to visit the Soviet Union and then selected to join the observation tour to the United States. Chern reached China in the spring of 1946, just as Hua was getting ready to leave for the United States, and the two of them met in Shanghai. Chern recalled that although Hua was on a mission, the two of them still talked a lot about mathematics, and gradually their mathematical interests grew closer.

Two Giants at a Distance

Hua's visit to the Soviet Union, which we have just touched upon, was familiar to everybody in China's intellectual circles, because he wrote a diary of some 30,000 words that was serialized across four issues of the Shanghai weekly magazine *Shi Yu Wen*, which was very popular in the 1940s. It is clear that in various periods in modern China, such legendary figures as Hua Luogeng attracted quite some public attention. In the Soviet Union, Hua met with Vinogradov, with whom he had long been familiar. A few years ago, Xu Lizhi, a favorite student of both Hua and Chern, remarked that both of his two mentors were worldly people, that is to say, they took an interest in politics. On the other hand, Xu Lizhi considered Xu Baolu, one of the "three outstanding heroes" of Southwest Associated University, to have been altogether out of this world.

Xu Baolu was born in Beijing in the same year as Hua; his ancestral home was in Hangzhou, his grandfather had been the prefect of Suzhou, his was the salt commissioner of Zhejiang province, and his brother-in-law was a famous redologist.[3] After graduating from Tsinghua University, Xu Baolu studied at University College London, where he earned his doctorate before returning to China and becoming a professor at Southwest Associated University. He is recognized as the first Chinese mathematician to gain recognition for work in mathematical statistics and probability theory. Unfortunately, he died young, during the Cultural Revolution, just over a year after Chern returned to China for his first visit. Xu Lizhi recalled that Xu Baolu was indifferent to fame and wealth, and did not care about power or official positions; this was a person who specialized in learning and was very noble but also liked to discuss politics. The personality of Xu Baolu is closely related to his background, and there should be some relationship between knowledge and physical condition. He never married, unlike Chern and especially Hua, who had many children.

During his time at Princeton, Hua completed a lot of excellent work in algebra, especially in classical group theory and division ring (infinite-dimensional algebras). He obtained an important result about semi-automorphisms, which Artin referred to as Hua's theorem, and gave a simple and direct proof of what later generations called the Cartan-Brauer-Hua theorem, which states that if the unit group of a proper division subring of a division ring is a proper subgroup of the unit group of the larger ring, then the subring is contained in its center. One of his American colleagues remarked that Hua possessed an uncanny ability to capture the best work of others and point out where those results could be improved. Weil commented that Hua played with matrices as if they were integers. In 1948, Hua was hired as a professor at the University of Illinois Urbana-Champaign, with an annual salary of more than 10,000 USD.

Hua brought his wife and three sons to the United States, but his eldest daughter, who was already in college and nurtured ambitions in progressive politics, stayed behind in China, and his youngest daughter, who had just been born, was taken back to her hometown in Jintan by her grandmother. That year, the Academia Sinica announced its first batch of academicians, and both Hua and Chern were on list, alongside three other mathematicians: Jiang Lifu, Xu Baolu, and Su Buqing. The University of Illinois Urbana-Champaign was famous for number theory, and Hua supervised two doctoral

---

[3] A scholar of the *Dream of the Red Chamber*, one of the four great classical novels of China (Translator's Note).

11 Hua Luogeng and Shiing-Shen Chern: Two Contemporary Chinese...

Institute of Advanced Study(photograph by the author, Princeton)

students in this discipline, one of whom, Raymond Ayoub, wrote an influential introduction to analytic number theory. In 1985, the year that Hua died, Ayoub announced that he had proved that irrationality of the Euler-Mascheroni constant, but he turned out to have been mistaken. This problem had a long history; Hardy once said that if anyone could prove it, he would give him his professorship at Cambridge.

On the last day of 1948, the year in which Hua arrived in Illinois, Chern also took his family from Shanghai and boarded a Pan American Airways flight to the United States. More than a month earlier, Chern had received a telegram invitation from J. Robert Oppenheimer, director of the Institute for Advanced Study in Princeton, and he made the decision to accept it after learning the Kuomintang government in Nanjing was on the point of collapse. Since his return to China more than a year earlier, Chern had been busy preparing to establish an Institute of Mathematics at Academia Sinica. Both before and after its establishment, in his role as its acting director, Chern recruited a wide range of young people, including such talents as Wu Wenjun, Liao Shantao, Zhou Yulin, Cao Xihua, Yang Zhongdao, and so on. Chern himself taught topology, and during this period he declined formal invitations from Princeton, Columbia, and other institutions, including the Tata Institute of Fundamental Research in India.

After his arrival in Princeton, Chern led a seminar and prepared lectures on geometry. During the summer, he taught at University of Chicago, where his friend Weil was a professor. Interestingly, another Professor there, Ernest Preston Lane, was supervising the doctoral candidate Sun Guangyuan, whom Chern had advised as a master's student. There were also two budding Chinese physicists at the University of Chicago, which later would give rise to a large number of economists: Chen-Ning Yang, who had just completed his Ph.D., and Tsung-Dao Lee, who was still working on his. Although Hua and Chern were both teaching in Illinois and should have had many opportunities to meet, Chern recalled that they seemed to have met only once when the University of Chicago invited Hua to present some lectures and then to bid their farewells before Hua's departure.

As autumn approached, with the birth of New China and its capital in Beijing, the Chinese mathematical community was in danger of losing two leader figures at the same time. Fortunately, a year later, Hua decided to abandon his high salary in the United States and move back to China with his family. As for his reasons, in spite of all kinds of suspicions and analyses, he was full of enthusiasm to serve his homeland, and his decision came as good news to the mathematical community in China. Many years later, Atle Selberg, a Norwegian-born American number theorist and recipient of the Fields Medal remarked that it is hard to imagine what would have happened in Chinese mathematics if Hua had never returned to China. Chern opted to stay in the United States, where he became an iconic figure for Chinese mathematicians in that country. His further contributions to Chinese mathematics would have to wait until his retirement.

Although the philosophy of the Chinese people, with the golden mean at its center, also has such sayings as "battle between shining lights" and "one mountain cannot accommodate two tigers," Hua and Chern maintained a lifelong friendship, not a close one, but one which stood the test of time. Both the Academia Sinica and later the Chinese Academy of Sciences were presented with this problem when selecting a director: Chern and Hua were two most worthy candidates, but only one of them could hold the position. Fortuitously, Hua was visiting the United States or preparing to go abroad during the establishment of the Institute of Mathematics at Academia Sinica, and Chern had already settled there when the Chinese Academy of Sciences was established.

If it is necessary to pick one of the two to stay in the United States, it is the opinion of the author that Chern was better suited to it. In the first place, the United States was the most active center in his area of research, and, in the second, his exchanges and collaborations with foreign colleagues were closer and more extensive. Hua may have been less lucky, or perhaps it was due to his personality, but he mostly worked alone and rarely received help or support from foreign colleagues. In terms of his ability to adapt to life in China, Hua, whose station in life was lower at birth, may have been the better equipped of the two. The facts bear out that the impact on him of the vicissitudes of the political movements was relatively light compared to other intellectuals. Even with respect to academic research, Hua exhibited an extremely strong instinct for survival. Despite a severe lack of access to information and communications, he was able to achieve results recognized around the world.

On the Western Pacific Coast

Upon his return to Beijing, Hua first taught at Tsinghua University before falling suffering the baptism of the Three-anti Campaign and the ideological reform movement. A photograph of Hua with Chiang Kai-shek brought him a lot of trouble; but Hua was, after all, a luminary worth uniting with. Mao Zedong had hosted a banquet for him in the past, and in the end he survived the test smoothly. The mutual revelations however caused an insurmountable gap between colleagues. It was not until the following year, when the Government Administration Council decided that he should serve as the director at the newly formed Institute of Mathematics at the Chinese Academy of Sciences, that his mood began to brighten. It is worth a mention that the chairman of the Preparatory Office of the Institute of Mathematics had originally been Su Buqing, and Hua had been one of the four deputy directors.

Hua flourished at the Institute of Mathematics over the next several years. In terms of organizational work, Hua recruited talents from all over the country and mobilized dozens of accomplished young mathematicians. He was attentive to both basic theory and applied mathematics and established two special working groups, in differential equations and number theory, while encouraging the other personnel to delve deeply into their own directions. Hua also presided over the first congress of the Chinese Mathematical Society (after the founding of the People's Republic of China), which elected him to be its chairman, and established the journal *Acta Mathematica Sinica*, serving as its editor-in-chief. Hua paid a visit to the Soviet Union with a delegation from the Chinese Academy of Sciences. If not for the sudden death of Stalin, it was expected that his *Additive Theory of Prime Numbers* would be awarded the Stalin prize the following year.

In 1955, the Chinese Academy of Sciences established its academic division, with Hua among the first batch of academic members. Hua and his Institute of Mathematics were very effective both in terms of academic research and in terms of teaching. He personally organized two seminars, one an introduction to number theory and the other on Goldbach's conjecture, the first of which later became his mathematical masterpiece, *Introduction to Number Theory*. One of the fruits of the second seminar was Wang Yuan's proof of the $(3, 4)$ and $(2, 3)$ cases, where (a, b) means that every sufficiently large even number can be written as a sum of two odd numbers with a number of prime factors not exceeding a, b, respectively. The case $(1, 1)$ is almost equivalent to Goldbach's original conjecture:

> every even number greater than or equal to 6 can be expressed as a sum of two odd prime numbers.

Here it is worth mentioning that the second of these seminars attracted the graduate students of Professor Min Sihe of Peking University, among them my own advisor Pan Chengdong. At that time, the mathematics department at Tsinghua University had been disbanded due to reorganization of colleges and departments in higher education, and the best parts of it had transferred to Peking University. A few years later, Pan, who was already a lecturer at Shandong University at the time, proved the $(1, 5)$ and $(1, 4)$ cases. Chen Jingrun, who proved the $(1, 2)$ case, was brought on from Xiamen University by Hua Luogeng himself. Earlier, Chen had written a letter to Hua about some of his achievements, and he admired Hua tremendously. Xu Chi discussed some of the things that happened during and after that period in his famous article on Chen Jingrun. To this day, Goldbach's conjecture remains unproven, and nobody has surpassed Chen's result.

In addition to number theory, Hua also made important contributions to algebra and function theory, in particular in the fields of classical groups and the theory of functions of several complex variables. The talents and assistants he trained in these fields included Wan Zhexian, Lu Qikeng, and Gong Sheng. His work *Harmonic Analysis of Functions of Several Complex Variables in the Classical Domains* earned Hua a 1956 first prize Natural Science Award under the President of the Chinese Academy of Sciences Guo Moruo, an award later regarded as equivalent to the National Natural Science Award. Many years later, his disciples Chen Jingrun, Wang Yuan, and Pan Chengdong won the same honor for their work on Goldbach's conjecture. Hua also discovered a group of differential operators with properties similar to those of harmonic operators, which later came to be known internationally as Hua operators.

11 Hua Luogeng and Shiing-Shen Chern: Two Contemporary Chinese...

Hua Luogeng, obsessed with the kingdom of mathematics

Also under Hua was a group of mathematicians working in other directions, most prominently Wu Wenjun and Feng Kang, who achieved world-renowned results in topology and computational mathematics, respectively. Wu Wenjun had been outstanding already when Chern had led the Institute of Mathematics of Academia Sinica. Later, he studied in Paris and obtained his doctorate there before returning to Beijing. His work on characteristic classes and imbedding classes earned him a first prize Natural Science Award in the same year as Hua. On the other hand, Feng Kang, one of the inventors of the finite element method, had been continuously engaged in research in China apart from 2 years of further study at the Steklov Institute of Mathematics in the Soviet Union. It was following the advice of Hua that he shifted from pure mathematics to research in computational mathematics, later becoming a deserved academic leader in this field. Four years after his death, he was posthumously awarded a first prize Natural Science Award for his *Symplectic Geometric Algorithms for Hamiltonian Systems*.

In China in the 1950s and the 1960s, it was impossible to steer clear of political activity, especially for Hua, who was a passionate person and liked to communicate. He had joined the Kuomintang as early as during his stint working at Jintan Middle School; during his time at Tsinghua University, he had actively participated in the December 9th Movement; at Southwest Associated University, he became close friends with the left-wing poet, Professor Wen

Hua Luogeng and his students

Yiduo. Later, Hua served for a long time as the leader of the Central Committee of the China Democratic League. In 1957, Hua, along with Zeng Zhaolun, Qian Jiaju, Tong Dizhou, and Qian Weichang of the China Democratic League, responded to the Hundred Flowers Campaign with a joint submission of several opinions on the reform of the scientific system to the State Council, unexpectedly causing a disaster for themselves. The chemist Zeng Zhaolun took the initiative to take responsibility for it, and Hua later admitted to mistakes in the newspaper. He and the economist Qian Jiaju (who nevertheless later suffered a worse fate) avoided the rightist label. Zeng Zhaolun, who was outspoken and righteous, came from the famous Zeng Guofan family of Hunan province.

Following the Anti-Rightist Campaign, next came the Great Leap Forward. Hua proposed that China should catch up to the United States with respect to 12 mathematical problems within 10 years and that all mathematical problems that have to do with computing technology, artificial satellites, large dams, and so on should be included. Obviously a mathematician as great as Hua was boasting against his will to say such things, but under the circumstances of the times he was still not considered to be "advanced" enough. Some young people in the Institute even suggested that it should take only 2 years to catch up with the United States in the area of partial differential equations. Once again, Hua was classified as a "conservative." Alongside consideration of his experiences in Old China and overseas, his repeated applications to join the Communist Party were rejected from within the Institute and the Academy of Sciences.

At that time, China had cut off ties with the West. In 1954, 1958, and 1974, Hua received invitations to present 45 min reports to the International Congress of Mathematicians but had to decline because he could not obtain government approval. In light of the situation and his advancing age, Hua turned his attention to applied mathematics on the eve of the Cultural Revolution. He devoted most of his energy in his later years to generalizing the critical path method and optimization method, with good results, which also ensured his relative safety during the "10-year catastrophe." But when Hua had to return to Beijing for hospitalization due to his first heart attack, he quietly thought about Goldbach's conjecture. He proposed an idea and line of thought, hoping that Wang Yuan and Pan Chengdong would collaborate with him, but received no response, because they had already secretly tried this approach.

On the Eastern Pacific Coast

While Hua was still vigorously leading the way for mathematics in China, Chern was wholeheartedly studying geometry in the United States and gradually making progress. In the summer of 1950, the International Congress of Mathematicians convened at Harvard after 14 years, and Chern was invited to present a 1-h lecture. His topic was the *Differential Geometry of Fiber Bundles*. Twenty years later, when the International Congress of Mathematicians convened in Nice, France, Chern delivered another 1-h lecture, entitled *Differential Geometry: Its Past and Future*. In these years, Chern was a glorious spokesperson for modern differential geometry. It had not been this way when he first arrived in the United States, however. At that time, the subject was considered a dead end, and was not even included in university curricula. Chern employed the essential elements of topology, critical points, the theory of fixed points, fiber bundles, and characteristic classes, to form global differential geometry.

During his 10 years in Chicago, Chern revived American differential geometry and established the American school in this field. His next destination was the coastal climate of the University of California, Berkeley, where his presence contributed to the rise of this public university from fourth to first place in mathematics in the United States. He improved the academic standing of the school in both geometry and topology. Chern worked closely with his many colleagues, including Phillip Griffiths, later the head of the Institute

Shiing-Shen Chern in 1976

for Advanced Study, and the legendary figure James Harris Simons. The collaboration between Chern and Griffiths was mainly in network geometry and exterior differential geometry, and because of Chern, Griffiths later visited China many times. He also served as the Secretary General of the International Mathematical Union. During his tenure, the International Congress of Mathematicians, affiliated with the International Mathematical Union, was successfully hosted in Beijing.

Chern and Simons collaborated to obtain the theory of Chern-Simons invariants, still an important area of research in theoretical physics, which was used by the physicist and recipient of the Fields Medal Edward Witten in his research into quantum field theory. Later, Simons became the head of the mathematics department at the University of New York, Stony Brook, where he worked with the physicist Chen-Ning Yang, and as a result, Yang realized at last after a lecture that the mathematical counterpart to the gauge field theory he had established with his collaborator Robert Mills was precisely the theory of fiber bundles established by Chern. In fact, the latter had appeared 10 years earlier than the former. In this way, modern geometry and modern physics are broadly and closely linked, indicating the academic significance of both fiber bundles and gauge field theory.

After achieving great fame in mathematics, Simons abandoned academics and went into financial investing, with equal success. In the spring of 2003, he chartered a private plane to visit Chern in Tianjin, with the landing application submitted by Chen-Ning Yang. In the context of the financial crisis sweeping the world, as the president of Renaissance Technology, Simons surpassed in one fell swoop the financial giant George Soros to obtain the first rank among global hedge fund managers, at the same time entering the Forbes Top 100 list of the richest people in the world. He injected funds toward the support of mathematical activities, including the academic conference held in Nankai to celebrate the 80th birthday of Shiing-Shen Chern.

Chern supervised 31 doctoral theses for the University of California, Berkeley, the most famous and accomplished of whom was Shing-Tung Yau, who later won the Fields Medal and solved many significant problems, such as the Calabi conjectures and the positive energy conjecture. During his time in Berkeley, Chern was also elected to the National Academy of Sciences and earned the Wolf Prize, for lifetime achievement in mathematics, specifically for his outstanding contributions to global differential geometry, which have influenced all of mathematics. The Wolf Prize was personally awarded by the President of Israel, and another honor awarded to Chern, the US National Medal of Science, was bestowed by President Ford at the White House. In 2009, the International Mathematical Union announced the establishment of the Chern Prize, to be awarded in person by the head of state of the host country of the quadrennial International Congress of Mathematicians.

Chern had not served in any administrative position after stepping down as acting director of the Institute of Mathematics of Academia Sinica and leaving China. His magnanimity in his interactions and his organizational skills, including in academic collaboration and guidance, nevertheless deeply impressed his American colleagues. Before a general election of the American Mathematical Society, Chern was asked if he would serve as president, but he firmly declined, serving instead as vice president for 2 years. Entering his 60th year, nostalgia for his hometown arose suddenly, and Chern took his wife and daughter back to his homeland, from which he had long been absent. He received an elite reception and also met with Hua Luogeng, who had been promoting his "double method" elsewhere when a telegram called him back to Beijing. What kind of scene was it? During the long years of the Cultural Revolution, two families shared a roast duck.

Chern came to a turning point in his life in his 70s, after he had retired from the University of California. In the spring of that year, he was in discussions with the leaders and his old friends at his alma mater, Nankai University, to establish the Nankai Institute of Mathematics in preparation for his return. In the autumn, however, the Mathematical Sciences Research Institute (MSRI) was established at Berkeley. Chern, who was one of its founders, was appointed as its first director. He had to postpone his return to China to settle down, and it was only 3 years later, after his term expired, that he accepted the invitation to serve as the director of the Institute of Mathematics at Nankai University. It is the opinion of the author that Chern did not choose to collaborate instead with his other alma mater, Tsinghua University, because he did not want to compete with Hua in the same city, who was still serving as the director of the Institute of Mathematics of the Chinese Academy of Sciences.

During his tenure as the director of MSRI, Chern frequently looked for opportunities to return to China. He met on many occasions with national leaders such as Deng Xiaoping and used his personal influence in an earnest effort to improve the level of mathematics in China. For example, the International Symposium on Differential Equations and Differential Geometry inaugurated by Chern was held for 7 consecutive years and the annual academic conference at Nankai University for 11 consecutive years. Chern also suggested a summer postgraduate workshop and personally taught classes at it himself. He spared no effort in recruiting and nurturing talents such as Long Yiming and Zhang Weiping, the former of whom is the current director of the Nankai Institute of Mathematics. After Chern died, they were both elected as academicians of the Chinese Academy of Sciences, the only two mathematicians in that election (in 2011, Chen Yongchuan, who had also been looked after by Chern, was also elected).

The Mathematical Sciences Research Institute at Berkeley

Conclusion: Remembrances and Prayers

In the early summer of 1985, the second year after Chern had been appointed as director of the Nankai Institute of Mathematics, which still had yet to be opened, Hua was invited to visit Japan. He gave a speech at the University of Tokyo, reviewing the work he had done since returning to China in the 1950s, divided into four parts chronologically with the 1970s and 1980s devoted mainly to the popularization of mathematics. Perhaps due to this retrospective opportunity, Hua was overexcited the night before and barely managed to take some rest by taking sleeping pills. The next day he insisted on abandoning his wheelchair and delivered his report for more than an hour standing up. When he finally sat down to thunderous applause and was about to accept flowers from a lady, he slipped from his chair. A few hours later, the University of Tokyo Affiliated Hospital announced that Hua Luogeng's heart had stopped beating, and he died of a heart attack at the age of 75.

At this time, Chern was in Tianjin, busy with the upcoming establishment of the Nankai Institute of Mathematics. When he learned the sad news, he immediately called the relevant authorities in Beijing and asked to attend the scattering ashes ceremony but was told that no guests from outside the city were to be invited to Beijing. Hua had been the vice chairman of the National Committee of the Chinese People's Political Consultative Conference, and the memorial ceremony was of a very high standard. But the author believes that, as a mathematician, Hua would have wanted in his soul for his old friend and colleague, who had known him for half a century, to come to see him off. Just 2 years earlier, Hua had visited Los Angeles, and Chern drove from Berkeley, more than 400 km away, to meet with him, their last meeting. It was in that year that Hua was elected as a foreign academician of the American Academy of Sciences via the joint nomination and recommendation of Chern and Felix Browder, whose father had served as the general secretary of the Communist Party of the United States and who had served as president of the American Mathematical Society, as had his brother William, and who had been studying for his Ph.D. at Princeton when Hua was visiting there. Chern wrote an academic introduction for this nomination.

Chern lived for nearly 20 more years after Hua passed away. Although he was still thinking about major problems in differential geometry, such as the existence of complex structures on six-dimensional spheres, mostly he enjoyed the fruits of his mathematical life and used his influence and appeal to promote Chinese mathematics. He also assisted with the successful bid for the International Congress of Mathematicians to convene in Beijing in 2002. As his twilight years approached, Chern received various honors, including the first Shaw Prize for science, with a one million US dollar award, the Lobachevsky Medal from Russia (named after the founder of non-Euclidean geometry), and election as a foreign academician to the French Academy of Sciences. The Chinese Mathematical Society established the Chern Shiing-Shen Mathematics Award (the Hua Luogeng Prize had already been established), and the new main building of the MSRI in the United States was named the Chern building.

Throughout Chinese history, the political status of mathematicians has always been relatively low. Before the twentieth century, it was rare for mathematicians to be received by the emperor, some exceptions being Qin Jiushao and Li Ye, two of the "Four Great Masters of the Song and Yuan Dynasties," both of whom lived in the thirteenth century. The former once lived in Lin'An (now Hangzhou), capital of the Southern Song dynasty. Zhao Yun, Lizong of the Song dynasty had summoned him to court; the latter was summoned by Kublai Khan, founder of the Yuan dynasty in Dadu (now Beijing). There was also the Qing dynasty mathematician Mei Wending, who was received three times on a dragon boat by Emperor Kangxi during his tour of Jiangnan in 1705. These three emperors were all either suffering through hardship or were foreign intruders. Among them, the only one who had done some research into mathematics should be Kangxi. Unfortunately, although Leibniz had sent a letter to him recommending that he establish a Beijing Academy of Sciences, there was no movement on this front.

The situation of Hua Luogeng and Shiing-Shen Chern in the twentieth century was different. Hua had received the courtesies of the heads of different political parties, including Chiang Kai-Shek, Mao Zedong, Hua Guofeng, Hu Yaobang, and so on, while Chern received honors transcending national boundaries: in addition to many individual meetings with Deng Xiaoping and Jiang Zemin, he was also decorated by the presidents of the United States and Israel. In fact, such honors had only been enjoyed by a few mathematicians throughout the world in all the history of mathematics, such as Leonhard Euler in the eighteenth century. However, the two of them had different mentalities with respect to political leaders. Hua was more like someone from past times, with a fearful side, whereas Chern was comfortable in every situation. This difference also appears in comparing the free verse that Chern composed as a boy and the ancient poems that Hua later exchanged with Mao Zedong. This difference was due to their different origins, experiences, educations, and environments, which also informed the differences between their academic paths and research styles.

11 Hua Luogeng and Shiing-Shen Chern: Two Contemporary Chinese...

Hua Luogeng and his wife pose for a group photograph with Shiing-Shen Chern and his wife (Beijing, 1972)

It is a pity that although Chern received the influence of famous schools in both the East and the West, he only spent so much of his life busily, and did not pause to think deeply about philosophy, as had the great mathematical ancestors of those who trained him in Paris, where he stayed for a time. The humanistic tradition of French mathematics extends from René Descartes to Henri Poincaré, the pioneers of geometry and topology, respectively, both of whom were also philosophers. As a result of this, France produces a mathematical luminary of the highest order about every 8–10 years. In China, we are more reliant on the sudden appearance of geniuses, which is especially clear in the case of Hua Luogeng; and Chern in any case did not obtain all of his education in China. On the occasion of the hundredth anniversary of the births of Hua Luogeng and Shiing-Shen Chern (and also Xu Baolu), while we were remembering and commemorating them, we were also sincerely praying for another Hua and another Chern to appear as soon as possible.

Among a number of precious historical materials released in the new century, I was fortunate enough to read two letters of recommendation written by the German mathematician Hermann Weyl for Hua in his early years. One of these was written in March 1943 to be delivered to Einstein and his colleagues at Princeton. Weyl believed that Hua and Chern were the two most outstanding mathematicians in China and particularly admired the work the former had done in analytic number theory and symplectic geometry. Weyl obtained a subsidy that year and invited Hua to come to Princeton to collaborate with his compatriot Carl Ludwig Siegel, but Hua declined. According to Wang Yuan, Hua may have worried about losing his originality. The second letter was written in March 1947 to S.S. Cairns, head of the mathematics department at Syracuse University in New York. By this time, he had had personal interactions and contact with Hua and believed that Hua was explosive and extraordinary quick in his work. Weyl wrote of Hua: "It makes him a very stimulating man to have around. He is cooperative and communicative and has a pleasant personality. We are all fond of him here and consider him a member of our group."

It is gratifying that, in comparison with the intractable grievances between individual Chinese physicists, Hua Luogeng and Shiing-Shen Chern spent their lives peacefully, and their friendship continued more or less unbroken through the years. This was the good fortune of the two of them and even more so that good fortune of Chinese mathematics. It was precisely due to them that Chinese mathematics finally took decisive steps to catch up with the times, after lagging behind the West for seven centuries. This also provided us with a necessary rise in confidence. Of course, their success also relied upon the pioneering work of senior mathematicians and educators, such as Jiang Lifu and Xiong Qinglai. With the continued improvement of its economic strength, China has the basic material support to catch up with the mathematical powerhouses of the world. The future will be brighter still if the humanistic environment of academia improves and becomes more conducive to bringing talent to the forefront.

Spring and Summer, 2009, Hangzhou and Hong Kong

12

"My Life Can Be Said to Form a Circle"—An Interview with Nobel Laureate Professor Chen-Ning Yang

Yang Chen-Ning spent his childhood at Tsinghua Campus in Beijing, and it was to Tsinghua Campus that he returned in 2003 when he came back to China, taking up residence in a small courtyard house in Shengyin Yard. He named this little home *A Return to Roots* and wrote a poem with the same title.

Mr. Yang and his wife, Weng Fan, spend most of their time in Tsinghua Campus, sometimes visiting the Chinese University of Hong Kong on holidays. In addition to attending necessary public events, he and Weng Fan have many mutual friends in their daily lives, and from time to time they attend private gatherings. I remember that Weng Fan once revealed to me that Mr. Yang likes to sleep late in the morning. A year ago, Mr. Yang was fitted with crutches, just in order to make walking a bit faster, safer, and more convenient.

Physicist Chen-Ning Yang

Chen-Ning Yang at Liuzhuang, West Lake, Hangzhou (photograph by the author)

I have been corresponding with Mr. Yang for nearly 3 years, starting when I invited him as a guest speaker to the Lecture Hall of Science at Zhejiang University, at which I serve as forum leader. After having been introduced through a friend, we corresponded by email; I wrote in Chinese and he replied in English, usually within an hour or two. Mr. Yang had originally promised to come to Hangzhou in the spring of 2013, and everything was arranged, but the morning before he left, Mr. Yang called me to cancel the trip because CCTV News had reported the presence of bird flu in Shanghai. This was during recess and his voice was so loud that the students could hear him. A few minutes later, he called again and said that there were no occurrences in Hangzhou yet, so that he could still come. But later that same day, it was reported that there had been a bird flu death in Hangzhou, and Mr. Yang wrote to me that night to confirm the cancellation of the trip. A few weeks later, I sent Mr. Yang my questions for this interview, and after 2 days I received his answers, which he had written in the margins himself with a fountain pen.

In March of this year, Mr. Yang informed me that he was coming to Hangzhou, and I finally had the opportunity to welcome him here. It was a privilege to spend a few hours alone with him at Liuzhuang West Lake State Guesthouse, and we continued our conversation during the trip to and from Zhejiang University and to the airport, even waiting in the VIP lounge. When I finished writing it up, I sent it Mr. Yang for several corrections.

From this trip to Hangzhou, we can see that although he is 93 years old, Mr. Yang still has a quick mind and a good memory; only the response of his right ear is slightly slow since he began wearing a hearing aid. During his speech at Zhejiang University, a reporter noticed that he didn't take a sip of water for 2 h. Mr. Yang is good at communicating with people and was happy to respond to requests from his admirers for group photos; during our time together, this included the waitstaff at Liuzhuang and in the VIP lounge at Hangzhou Xiaoshan Airport:

1 Cai: Hello, Mr. Yang! First of all, thank you very much for accepting our invitation to be our guest at the Science Lecture Hall of Zhejiang University and also for agreeing to this interview. Unfortunately, President Yang Wei, who wrote a letter to you himself, has left Zhejiang University and cannot be here to welcome you in person. When he learned that you were coming to Zhejiang University, he wrote back to me and said, "I am very glad that Mr. Yang is finally able to come!" I would like to know how many times you have been to Hangzhou before, and do you remember when you first saw West Lake?

Yang: I came to Hangzhou for the first time in the summer of 1972, and I have been here five or six times. Almost every time I came to Zhejiang University; the first time was during the Cultural Revolution; I visited Yuquan campus for a stroll, but did not encounter a single acquaintance. President Yang Wei is also a Tsinghua alumnus, but we haven't met since he left Hangzhou to work in Beijing. (*Note: The day before Mr. Yang's visit to Zhejiang University, he had already met with the newly appointed President Lin Jianhua. On the day of the lecture, the future President Wu Chaohui personally accompanied him. When chatting with me, Mr. Yang mentioned the late professors Prof. Wang Rong and Prof. Li Wenzhu of the Physics Department of Zhejiang University and asked if Mr. Tsung-Dao Lee had recently returned to pay any visits to his alma mater*).

2 Cai: You were born in Hefei, the same town as Li Hongzhang (exactly one century apart). At that time, Hefei was just a county in Anhui, and your father was a high school teacher in Anqing, the provincial capital, which was then called Huaining; this is where your name Chen-Ning comes from. Also, your childhood friend Deng Jiaxian was born in Huaining; he was your classmate through high school and college and later a roommate in New Jersey when you were studying in the United States. Are you familiar with the poet Hai Zi, born in 1964 in a village in Huaining county, who committed suicide at the age of 25 in Shanhaiguan and is now practically a household name in China? Have you read any of his poems.

Yang: I haven't heard of Hai Zi, nor have I read any of his poems. I was born on a small street called Si Gu Xiang in Hefei, where I lived until I was 6 years old. Some years ago I went back to Hefei and visited the "Former Residence of Yang Zhen-Ning," but it was not the same place where I spent time as a child. Of course, I didn't say that to the receptionist. Si Gu Xiang was named after four ancient tombs that once stood there,[1] and it is said that the name of this alley was recorded in the *Hefei County Annals* more than 200 years ago.

3 Cai: Your father left to study in the United States when you were less than a year old. You saw him again at the Port of Shanghai when you were 6 years old, and your family went to Xiamen together, where you saw electric lights, ate bananas, and drank milk for the first time. A year later your father was hired by Tsinghua University, and you came to Beijing and lived at Tsinghua Campus for 8 years. It is said that you excelled in mathematics as a child and could already read Hardy's An Introduction to the Theory of Numbers *and E. T. Bell's* Men of Mathematics. *However, your father, himself a mathematics professor, hired a teacher of ancient languages to teach you Mencius. How did this experience shape your later life?*

Yang: My father taught at Xiamen University for 1 year after he returned to China, and in the summer of 1929, he accepted an offer from Tsinghua University, and our family went from Xiamen to Beijing via Shanghai. We lived at Tsinghua Campus in Beijing, which was called Beiping at that time. My 8 years at Tsinghua Campus were beautiful, and everything is very nostalgic for me. At that time, Tsinghua University was small, but there were about 50 children of faculty members, so an elementary school was established and I studied there. The stories in Mencius had a great influence on me, showing me the worldview of traditional Chinese culture and the principles of being human.

4 Cai: My father was 1 year older than you and, like you, attended National Southwestern Associated University in Kunming in the 1940s. He studied history, and, also like you, he attended Mr. Wen Yiduo's poetry class; but he passed away 34 years ago. I would like to ask you how many students were enrolled at Southwest

---

[1] *Si Gu Xiang* means *Four Ancients Alley* in Chinese (Translator's Note).

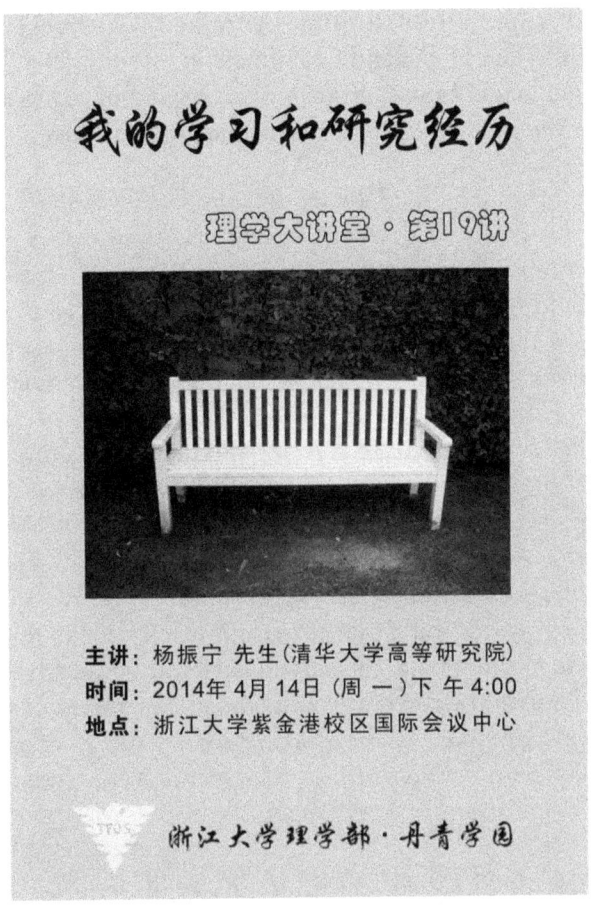

Zhejiang University School of Sciences Lecture Hall Poster

Associated University each year, and how many professors and teachers were there? How was the life of the students? Was there any exchange between Southwest Associated University and Zhejiang University, which was also located in Guizhou in Southwest China during the war? What is your fondest memory there?

Yang: At that time, the university enrolled about 400 new students each year, and I don't remember any exchanges with Zhejiang University. The undergraduates of National Southwestern Associated University belonged to three universities that had been merged into one, with graduate students managed separately by each school, although at that time it seemed that only Tsinghua had a graduate school, on account of the Boxer Rebellion Indemnities. As for my personal hobbies, I liked to sing when I was young, although I didn't sing very well. After the Marco Polo Bridge incident, our family first returned to Hefei, and I continued my studies. The following year, I had not yet graduated from high school and was admitted to the Southwest Associated University with the same academic qualifications.

5 Cai: Your father, Mr. Yang Wuzhi, was the first Chinese Ph.D. in number theory (at the University of Chicago), a senior colleague whom I admired. He proved that every positive integer admits a representation as a sum of certain cubic polynomials, which Mr. Wang Yuan later praised as a very good result for those days. In fact, this was a variation of Waring's problem, and he proved that each positive integer can be represented as a sum of at most nine tetrahedral numbers. Eighty-six years have passed, and this has been improved only to eight tetrahedral numbers. In 1994, you published a paper in Science in China Mathematics *(with Yuefan Deng) that argued heuristically and with numerical data that every positive integer should admit a presentation as a sum of at most five tetrahedral numbers and sufficiently large positive integers as a sum of at most four tetrahedral numbers. Was this a tribute in memory of your father?*

Yang: I cannot say this was a tribute to my father. I tried to read his doctoral dissertation but found that I couldn't understand it quickly because there were so many lemmas. I figured it would take at least a week or two to understand it, so I gave up. (*Note: Coincidentally, the day Mr. Yang gave a lecture at the Zhejiang University was April 14, 2014, which also happened to be the 118th anniversary of the birth of Mr. Yang Wuzhi.*)

6 Cai: When your father was teaching at Tsinghua University, he sparked Hua Luogeng's interest in number theory, and after Hua returned from a visit to England, your father advocated for his extraordinary promotion to full professor with only a secondary school education. Were Hua Luogeng, Shiing-Shen Chern, and Pao-Lu Hsu already famous when you entered National Southwestern Associated University? I heard that your first girlfriend was a senior student of mathematics and had been your father's teaching assistant. What I want to know is why you entered the chemistry department instead of the mathematics department. There is a legend that at that time you felt that there was no Nobel Prize in Mathematics?

Yang: When I was a student at National Southwestern Associated University, Hua, Chern, and Xu were already very famous. I had liked Zhang Jingzhao, who was from Shengzhou, Zhejiang, and later taught at Peking University. As for the legend, it is completely unfounded. I chose chemistry because my father thought that chemistry might be more useful than mathematics. But I didn't even wait for the start of the school year to petition Wu Youxun, Dean of the Faculty of Science, to switch to the Physics Department, in which I was successful. In those days, all the girls wore blue cloth coats, but Jingzhao Zhang stood out in a red suit. (*Note: Mr. Yang once said that before he met Zhang Jingzhao, his mood was like a calm lake, and after he met her, it turned into a storm; so perhaps this was his unrequited love. In 1968, in the midst of the Cultural Revolution, Zhang Jingzhao committed suicide in a restroom at Peking University.*)

7 Cai: In 1945, you went to study in the United States on an American troop carrier via India. That was your first trip abroad, right? Do you remember that trip? How long did it take? What ports did you pass through? I remember that when Hua Luogeng set off from Kunming to the Soviet Union in 1946, he also traveled from Calcutta. He chose a combination of land and air routes, passing through Pakistan, Iraq, Iran, Azerbaijan, Georgia, and other countries, and it took more than a month.

Yang: About 20 of the United States Boxer Indemnity students flew to Calcutta from Kunming at the end of August 1945, boarded an American troop carrier in late October, and arrived in New York in late November via the Red Sea, the Mediterranean Sea (and the Atlantic Ocean). We did not take a ship across the Pacific Ocean because, although Japan was defeated at that time, it had not yet officially signed the surrender, and although the sea route was shorter, it was more dangerous.

8 Cai: Chicago, on the shores of Lake Michigan, was at the center of American culture in the nineteenth century and into the first half of the twentieth century. The magazine Poetry, *founded in 1912, is considered the most important venue for avant-garde poetry in the twentieth century;* Sister Carrie *by Theodore Dreiser broke new ground for American literature, and Ernest Hemingway was born outside Chicago. Later, Saul Bellow taught at University of Chicago. The University of Chicago is also world-renowned for its mathematics and physics, and you went to Princeton, the world's highest temple to natural science, after earning your doctorate at the University of Chicago. You have done major work throughout your life in the glow of both these institutions. What are your different feelings and memories of these two cities? Did the Institute for Advanced Study contribute more to mathematics than to physics?*

Yang: I learned the methods and attitudes of doing research at the University of Chicago, and then my 17 years at Princeton were the most successful 17 years of my life in research. But you are right that the Institute for Advanced Study has done more in mathematics than in physics. (*Note: Before Mr. Yang came to Hangzhou, he saw an article on von Neumann in my book* Unattainable

Heights: The Shining Stars of the Mathematical Sky, *published in Taipei, in which I mentioned that when the Institute for Advanced Study celebrated its 60th birthday in the 1990s, it commemorated three landmark achievements: Gödel's research on the continuum hypothesis, von Neumann's research on algebra and the mathematical foundations of quantum mechanics, and the work of Chen-Ning Yang and Tsung-Dao Lee on parity violation. Yang was not yet aware of this and wrote to ask me for the source in order to verify it for himself.*)

9 Cai: You have said that although Newton knew that his Mathematical Principles of Natural Philosophy *was an extremely beautiful work, he could not have realized that his work would change mankind's understanding of the basic structure of the physical and biological world and would change forever the relationship between mankind and the environment. In this sense, how would you rate your work? Examples include non-abelian gauge theory, the theory of parity violation, and the Yang-Baxter equation. Is non-abelian gauge theory dominant in gauge theory? What is the current status of research on gauge theory, which along with Maxwell's electromagnetic field theory and Einstein's gravitational field theory are the three major field theories discovered by humans so far?*

Yang: I was fortunate enough to realize early on that there had to be a mathematical basic theory or principle controlling the propagation of the "force." At the same time, I became interested in symmetry very early on, and the two together gave rise to non-abelian gauge theory. This theory is obviously an important step, but it does not yet fully address the ultimate goal of unified field theory. This ultimate goal was also the goal to which Einstein devoted himself in his later years, when he tried unsuccessfully to establish a unified field theory that would encompass both electromagnetism and general relativity.

10 Cai: In 2000, the existence and mass gap problem of Yang-Mills theory was one of the seven Millennium Prize Problems *proposed by the Clay Mathematics Institute in New York, with your name and that of Mills the only names belonging to nonmathematicians in the list, something to be envied by many a mathematician. Do you think that your mathematical intuition comes from genetics or from other aspects of your education? Someone described the research style of Poincaré as that of a pioneer, not a colonist. What about your own style of research? Do you love whatever task you take to, not abandoning the old but creating something new?*

Yang: I think I appreciate mathematics partly because of genetics and partly because I had the opportunity to be exposed to it very early. As I said before, my initial understanding of group theory came from my father, and the bachelor's and master's theses I completed under Mr. Wu Dayou and Mr. Wang

Zhuxi, respectively, when I was in Kunming were on symmetry principles and statistical mechanics, which later became my main research directions for my entire life. In terms of research topics, I like to work on new things without blindly seeking novelty. For example, string theory is an emerging field in theoretical physics, and many mathematicians are involved in it, but it seems that string theory is not yet as significant to physics as it is to mathematics.

11 Cai: You have quoted Einstein as saying, "The true laws cannot be linear, nor can they be derived from linearity." You have also talked about Einstein's preference for geometry, and his suggestion that gravity and mechanics should be described by Riemannian geometry and that electromagnetism is also geometric. You also pointed out that the geometric structures for which Einstein was looking were gauge fields and that the simplest abelian gauge field was Maxwell's electromagnetic field, while non-abelian gauge fields are necessarily nonlinear. You were very excited when you discovered in 1975 that gauge fields are closely related to Mr. Chern's fiber bundles, and you should have a personal preference for geometry. However, in recent years, there have been many international academic conferences on physics and number theory, for example, the application of the Möbius inversion formula to a range of problems in condensed matter physics. Are you aware of the progress in this area?

Yang: This sounds very interesting, and if there are really results in this area, please write to me when you find any relevant articles or materials. In October 2011, I gave a report at the 100th anniversary of Mr. Chern's birth held by Nankai University. Later, I edited the speech into an article *Quantum Numbers, Chern Classes and a Bodhisattva* and submitted it to *Physics Today* for publication. I remember that I sent you a proof copy of it, and one of the subsections was entitled *When Physics Meets Geometry*. (Note: During his lecture, Mr. Yang mentioned that he had worked with Einstein for many years and often saw him at Princeton and knew his routine. Once Mr. Yang was waiting by the road when Einstein walked by, and he had his young eldest son stand in front of Einstein to have his picture taken. Later, Mr. Yang remarked that he still regrets that he did not take a picture with Einstein at that time.)

12 Cai: Nearly 12 years ago, you proposed four words as a tentative comparison between four theoretical physicists of the generation just before yours: Pauli (forceful), Fermi (robust, strong), Heisenberg (profound insight), and Dirac (Cartesian purity). You also mentioned that the relationship between Pauli and Heisenberg was once very tense. If you were to make an attempted comparison between yourself and a few of your peers, what words would you choose (Chen-Ning Yang, Tsung-Dao Lee, Chien-Shiung Wu)?

Yang: Sometimes physics and mathematics are different. Take the nineteenth century, for example, mathematics, viewed from a distance, had more than 20 large and small hills, while physics, viewed from a distance, had only a few large hills.

13 Cai: You were a speaker at the 1960 Gibbs Lecture of the American Mathematical Society, and at the 2008 Einstein Lecture of the American Mathematical Society, the British-American physicist Freeman Dyson praised you and your gauge theory, and his lecture, Birds and Frogs, *was later published in the* Notices of the American Mathematical Society *(2009); this article has been very influential among mathematicians. I have also read his book about you,* A Conservative Revolutionary, *in which he states that "Yang's sense of the beauty of mathematics illuminates all his work." He considered you the architect par excellence of twentieth-century physics after Einstein and Dirac. I noticed that some of your masterpieces were done in collaboration with others, which shows that you are good at communicating and collaborating with others. In his later years, Shiing-Shen Chern summarized his three closest Chinese friends and three closest foreign friends. If you were to make a similar summary, what would be your answer?*

Yang: I have done research with many people in my life, and the most successful collaborations have been with Tsung-Dao Lee, Mills, and Wu Dajun. Although the Yang-Baxter equations are also famous, Baxter and I did not work together, and I was unhappy about his fate. (*Note: Baxter worked independently on these equations after Yang. Wu Dajun was a student of Yang, a professor at Harvard University, and a member of* Academia Sinica *in Taiwan, and the two collaborated extensively in areas such as unified quantum field theory and particle physics.*)

14 Cai: Dyson mentioned that he learned more from talking to Fermi for 20 min at a critical point in his academic career than he did from his 20 years of association with Oppenheimer, father of the atomic bomb. After your doctorate, you worked as an assistant to Fermi for a year. On the 100th anniversary of his birth, you also wrote an article in his honor, holding him up as one of the most respected and admired of all the great physicists of the twentieth century. I would like to know what you would say about your mentor, Edward Teller, the father of the hydrogen bomb, and his Hungarian hometown friend, von Neumann, who like you, originally studied chemistry and worked across a range of fields like nobody else. Did you have any interaction with him at Princeton? I also noticed that although you have many collaborators, you have not mentored many students. Is it because you expected too much of your students?

Yang: Regarding Teller, he was very smart and extremely innovative. He was good to his friends, and I personally never saw him treating people badly. I was very involved in research when I arrived at Princeton, and by the time I had achieved my results, von Neumann was very ill, having been exposed to nuclear radiation as a result of his involvement in the atomic bomb tests. I did not have many Ph.D. students because I was reluctant to accept new graduate students when I don't have good topics for them. (*Note: Mr. Yang carried out an analysis of the differences in character between Deng Jiaxian and Oppenheimer, saying that the former was very reserved and the latter was sharp and that both China and the United States found the most suitable person to lead the design of the atomic bomb.*)

15 Cai: Not long ago, the Chinese government proposed that universities should lead culture, a fourth task in addition to the cultivation of human ability, scientific research, and serving the economy. In your opinion, what can and should university teachers, including those in the natural sciences, do? After the 1952 reorganization of Chinese higher education and the subsequent division of arts and sciences, this task has become significantly more difficult. Is this related to the exclusive reliance on Confucianism for more than two millennia? After the Duke of Zhou and Mozi, there was no one else who straddled the arts and sciences, and Confucius never seemed to mention mathematics or physics in his doctrine.

Yang: Every kind of work in introductory scientific research and popularization deserves attention and value around the world. I have one idea: right now Chinese college students are struggling with employment; if ordinary undergraduate colleges and universities, secondary schools, and other institutions had people dedicated full time to the popularization of science and teaching of scientific culture, then not only would it contribute to improving the scientific literacy of everyone, but we can also solve a large number of problems of unemployment among the educated youth.

16 Cai: After reading your book Dawn Collection *(SDX Joint Publishing Company, 2008), I realized that you are also a great writer of popular science, reminding me of Euler in the eighteenth century. In the article* Beauty and Physics *you describe three domains of physics as experimental, phenomenological, and theoretical and their relationship with mathematics, pointing out that the latter is the highest state of physics, with examples given by Tycho Brahe, Kepler, and Newton, respectively. You also mentioned that mathematics is the language of theoretical physics and compared the relationship between mathematics and physics to two leaves that overlap at the stem. In 2009, more than a dozen of us mathematicians from different universities at home and abroad founded the quarterly journal* Mathematical Culture *to explore the culture, ideas, and methods of mathematics and to create a bridge between mathematics and the natural sciences, humanities, engineering, and everyday life. Have you been able to receive regular copies of our journal?*

Yang: I read a few issues and thought it was very good. (*Note: During our casual conversation, Mr. Yang talked about the article introducing the number theorist, Min Sihe, serialized in the latest issue of* Mathematical Culture. *He remembered its details and added that he admired Mr. Min, and in fact it turned out that the Yang and Min families were closer during the National Southwestern Associated University period. This article was coauthored by editorial board member Dr. Zhang Yingbo and one of the editors-in-chief, Dr. Liu Jianya, and Mr. Yang also mentioned his first meeting with another editor-in-chief of Mathematical Culture, Dr. Tang Tao, in Australia.*)

17 Cai: In 1986, you talked about the contributions of the German mathematician Hermann Weyl to physics, about two things he cherished in his life—gauge field theory and non-abelian Lie groups—which later you and Mills synthesized to develop non-abelian gauge field theory. You also refer to Weyl's book Philosophy of Mathematics and Natural Science *(1926) and to the lines by T.S. Eliot that are quoted in the book, "Home is where one starts from. As we grow older / The world becomes stranger, the pattern more complicated/Of dead and living." This is a verse from* East Coker *(1943), one of the* Four Quartets; *do you know who translated it into Chinese?*

Yang: I don't know about that. When we were in Haining, Weng Fan and her sister had visited Xu Zhimo's former residence.[2] (*Note: During this visit to Hangzhou, Mr. Yang was invited to Haining, Jiaxing, to attend the commemoration of the 100th anniversary of the birth of Zha Jimin, a famous Hong Kong industrialist and philanthropist. During our chat, Mr. Yang also talked about his dialogue with Mo Yan, contact with the poet Bei Dao, and assistance for his return to China to visit relatives and work in Hong Kong. He also asked me how I knew Mo Yan and I mentioned Ha Jin and his novel* Waiting, *about which both he and Weng Fan remarked that it was difficult to read.*)

18 Cai: Qin Jiushao was a mathematician of the Song Dynasty and the most accomplished and internationally renowned ancient Chinese mathematician; the Chinese remainder theorem, which he discovered, appears in every course in basic number theory in China and abroad and has important applications in abstract

---

[2] Xu Zhimo was a Chinese romantic poet of the early twentieth century (Translator's Note).

algebra and a number of other current areas of science and mathematics, such as cryptography. Last year, a bridge named after him, the Dao Gu Bridge, was renamed and erected on the banks of the Xixi River in Hangzhou. George Sarton, the Belgian-American historian of science and founder of the modern discipline of history of science, wrote that "[Qin Jiushao was] one of the greatest mathematicians of his race, of his time, and indeed of all times." In your opinion, who was the best physicist in ancient China? And what was the best work?

Yang (*after thinking for a while*): Shen Kuo, who was from Hangzhou and wrote *Mengxi Bitan* (*Dream Pool Essays*), which talked about the discovery of optics. Do you know it? During the Qianlong period of the Qing dynasty, foreign missionaries brought with them glass, which emptied the streets of Beijing as everyone went to have a look, since at that time, the Chinese did not yet know how to make glass and women used bronze mirrors to look at their reflections. (*Note: I had planned to suggest on my way to see him off that Mr. Yang should stop at the Dao Gu Bridge, named after Qin Jiushao, and in this way have the two most accomplished scientists in Chinese history meet for a photo. Unfortunately, that day happened to be the first day the Hangzhou Airport Expressway was closed for road construction, so we were forced to take a detour.*)

The author interviews Chen-Ning Yang at Liuzhuang

19 Cai: Physics today faces a strong competitor, the "upstart" biology. Would you agree with this? The double helix structure of DNA discovered by Watson and Crick stands with Newton's law of universal gravitation in physics. In this comparison, Darwin's theory of biological evolution is equal to the law of free falling objects in physics, discovered by Galileo, although he failed to explain its precise cause. Has the Einstein of biologists not yet appeared?

Yang: You've heard the story of Rosalind Franklin, I guess, and if not I'll tell you about it. She was a British chemist and crystallographer, and in fact it was Miss Franklin who first took the double helix crystal diffraction pictures (somewhat blurred) that became the key factor in finally working out the structure of DNA. And yet, Franklin received no honors during her lifetime. She died of ovarian cancer in 1958 at the age of 37, and 4 years later, Watson, Crete, and Wilkins received the Nobel Prize in Physiology or Medicine for their work on DNA. Wilkins was a colleague of Franklin's at King's College, London. Watson and Crick came to Miss Franklin to collaborate, but she turned them down. She thought herself capable of obtaining clear pictures, which Wilkins had seen and described to Watson and Crick without her permission, enabling them to find the double helix structural model for DNA.

20 Cai: As I understand it, you also have a special respect for another, male, Franklin, an American who was both a scientist and a politician, and you named both yourself and your eldest son Franklin in English, although you never got involved in politics. Not long ago, Li Yuanchao, a mathematics graduate of Fudan, was elected vice president of the country. You met with Deng Xiaoping before, and his wife Zhuo Lin graduated from the physics department of Peking University (Liu Shaoqi's[3] wife Wang Guangmei also studied physics at Fu Jen Catholic University, where she stayed on to teach after completing her master's degree. She was a classmate of your first wife, Du Zhili), and a couple of their children also studied physics, so it seems that the Deng family physicists are more influential than the politicians (laughs). What do you make of all this?

(Note: Mr. Yang did not answer this question, but talked to me instead about the time when Mao Zedong met with him in his study in Zhongnanhai in 1973, mentioning one detail in particular. At that meeting, Mao talked to him mainly about philosophy. When the discussion came to Guan Zhong, a politician and military man during the Warring States period, he was particularly excited and suddenly sped up his speech, and Mr. Yang could not understand his Hunan accent for a while. At this time, Zhou Enlai, who was sitting on the side, took the

---

[3] A revolutionary and politician who held extremely high positions for 15 years under Chairman Mao prior to eventually falling out of favor (Translator's Note).

initiative to switch places with Zhou Peiyuan, a physicist beside Mao Zedong, and temporarily acted as an interpreter for Mr. Yang. Coincidentally, Liuzhuang, where Mr. Yang stayed during his trip to Hangzhou, was also the residence of Mao Zedong during his many visits to Hangzhou in his later years. Mr. Yang once pointed to the octagonal building next to his bedroom and told me that the teams of Zhou Enlai and Nixon negotiated and drafted the Shanghai Communiqué there.)

21 Cai: In the summer of 1971, when you returned to China for the first time after a gap of 26 years, why did you choose to enter from Myanmar? I remember that Dr. Kissinger's visit to China later that year was by way of Pakistan. You met your parents in Shanghai and visited places like Beijing and Dazhai in Shanxi. That trip must have been very emotional for you. Did you have a vision of the China of today at that time?

Yang: At that time, only Air France had flights to China and they had to go through Yangon, then the capital of Myanmar, so I had no other choice—I went to Europe. Yes, China has changed so much—I could never have imagined how much China would change in the next 40 years. That air trip was memorable, and I remember that the first thing I saw from the air after crossing the border was Kunming, the city where I had attended college and lived before I left China. However, starting from 1957, my father was able to reunite with me in Geneva.

22 Cai: There is a saying in Chinese folklore, the "seventh year itch."[4] You and Weng Fan have been married for 10 years now. How do you feel about married life? I heard that you like to use a DV recorder, so you must like to travel. How many countries have you traveled to in your life? How many cities have you lived in for more than a year? What places have you and Weng Fan traveled together since your marriage?

Yang: We have been to many countries over the years, but we only live in Beijing and Hong Kong. We have been to the United States many times, but only for a month or two at a time. (*Note: On a couple of occasions, Weng Fan interjected that her term for Mr. Yang was Darling, in English, which reminded me of Song Meiling and Chiang Kai-shek, a high-profile couple that attracted much attention during the Republican Era. We also spent some time going down the list of leading personalities in Hangzhou, and when it came to Internet hero Jack Ma, both Mr. Yang and Weng Fan mentioned that they had dined with him and discussed some interesting things.*)

---

[4] Indicating the end of the honeymoon phase of a relationship (Translator's Note).

23 Cai: *When you were 30 years old, you were about to embark on your research in gauge field theory, right? When you were 60 years old, you said that your father's friend, Mr. Zhu Ziqing, had changed Li Shangyin's poem "The sunset is infinitely good, only it's near dusk" to "But the sunset is infinitely good, why feel melancholy near to dusk?", and you thought this change was in line with your state of mind in your later years. I would like to tell you that a professor from the College of Humanities of Zhejiang University recently proved that this "only" actually meant "just like" during the Tang dynasty, so that Mr. Zhu Ziqing did not need to change this poem. I would like to know what your special feelings were when you celebrated your 19th birthday. Did you write a poem?*

Yang: Yes, I have written one, and it is about to be published. (*Note: I later read this poem, a free verse poem written in English, in the English edition of* Selected Papers of Chen-Ning Yang, Volume 2 (2013), *which was sent by Mr. Yang and published by World Scientific Press in Singapore, reproduced below*).

On Reaching Age 90

> Mine has been
> A promising life, fully fulfilled
> A dedicated life, with purpose and principle,
> A happy life with no remorse or resentment,
> And a long life …
> Traversed in deep gratitude.

24 Cai: *You returned to Beijing 10 years ago to settle there and directed the creation of the Institute for Advanced Study at Tsinghua University and served as its honorary director. Can you tell us about your daily life now and how you divide your time between Beijing and Hong Kong? How is Tsinghua Campus today different in your mind from the Tsinghua Garden of your childhood? I know that your parents are buried in Suzhou—do you still remember them from time to time? The photo of your reunion with your father in Geneva in 1957 is moving and unforgettable. Can you tell us your secret for taking care of your body, which is still healthy and able to travel and speak? Of course, no one can avoid the laws of nature. What will your epitaph be? Do you have a particular mathematical formula that you would like to use?*

Yang: I am close to my three younger siblings, and we certainly miss both parents and everything from our childhood at times. Tsinghua Campus is the place where I grew up as a child, and my life can be said to form a circle. I started from one place, traveled far away, and now I have come back. (*Note: In 2003, Mr. Yang returned to China. Later, a small courtyard was built in Zhaolan at Tsinghua Campus as a place for Mr. Yang to live and work. He named this small building* A Return to Roots *and wrote a poem with the same title. "Formerly responsible for a thousand searches, high on Jiuren Peak. Deep studies into symmetry, courage soaring to the clouds. The new heavens of our divine land have changed, the heavy mission of my homeland. Students aspire to the heights, I am the pine marking the trail. Three melodies of a thousand years, in the midst of talk and laughter. A new career late in years, the old man returns to the roots of his eastern fence."*[5]*).*

25 Cai: Next year is the 50th anniversary of the discovery of Yang-Mills theory. What commemorative events will you attend? When you first discovered that quantum physics revealed a striking relationship between elementary particle physics and the mathematics of geometric objects, this prediction was later confirmed at Brockhaven in New York, Stanford, the European Institute for Particle Physics in Geneva, and Tsukuba in Japan. How come China's high-energy physics laboratory didn't do it? How many more years do you think it will take to prove the existence and mass gap hypothesis? Can it be solved within this generation of mathematicians and physicists?

Yang: It's the 60th anniversary. Ten years ago there was a book published for the 50th anniversary, edited by Gerard 't Hooft. Chinese physics is growing quickly and the future is very bright. The pessimistic view of many people is because they do not know the inside story. (*Note: Gerard 't Hooft is a Dutch physicist born in 1946, known for his contributions to the development of gauge theory in particle physics and for his doctoral thesis introducing the technique of dimensional regularization and giving a proof of the renormalizability of Yang-Mills theory; he was awarded the Wolf and Nobel Prizes in Physics in 1981 and 1999, and he is also a foreign member of the French Academy of Sciences and the US National Academy of Sciences*).

---

[5] There is a pun on the family name of Weng Fan in the last line with the character 翁, meaning *old man* or *father* (as an honorific), impossible to render in English.

GPSR Compliance

The European Union's (EU) General Product Safety Regulation (GPSR) is a set of rules that requires consumer products to be safe and our obligations to ensure this.

If you have any concerns about our products, you can contact us on

ProductSafety@springernature.com

In case Publisher is established outside the EU, the EU authorized representative is:

Springer Nature Customer Service Center GmbH
Europaplatz 3
69115 Heidelberg, Germany

www.ingramcontent.com/pod-product-compliance
Lightning Source LLC
LaVergne TN
LVHW011001250326
834688LV00003B/51